WELL BEING

A Case Study on Mong Polyandrous and Monoganous Tribal People

WELL BEING

A Case Study on Mong Polyandrous and Monoganous Tribal People

By
R.L. Zinta
Sunder Kala Negi

Neha Publishers & Distributors
4832/24, Prahlad Lane, S-207 Ansari Raod, Daryaganj, New Delhi-110002

Published by
Neha Publishers & Distributors
4832/24, Prahlad Lane, S-207 Ansari Raod, Daryaganj,
New Delhi-110002
Tel.: 011-43570976, 23278261
Email-nehapubdistributors@gmail.com

First Edition : 2013

ISBN : 978-93-80318- 36-3

Price : 525/-

Laser Typesetting by : Vinod Bhardwaj

Printed at: Vikas Computers, Delhi

PRINTED IN INDIA

Published by Neha Publishers & Distributors, New Delhi-110002

PREFACE

Life as a love affair seems to be a natural instinct full of sacrifice and spiritually imbibed emotion that has one entrance with many exit. A perfect love definitely produces perfect citizen with balanced personality. The love before marriage is just infatuation and attraction of the face that becomes deeper and real after marriage through concordance and confidence. The Ishq is the deeper love that involves intelligence may become obsessive in later life. Love is neither disapproved by religion nor prohibited by the Law but for every heart is in God's hands. In the present scenario, it has converted into profession as well as sex tourism wherein the traffickers, escorters and touters are heavily involved for ostracizing, pimpling and bartering the females at every chokepoint of life. Little girls in the United States are encouraged to become "Sesame Street" characters (smart, athletic, and successful like boys) but in our country their picture is spunky and full of checkout. They are conditioned that the boys are from Mars engaged in hardware stuff whereas the girls from Venus made for love relationships. Despite this the marriage is the strong basis of a family structure and turning point in ones personal and social life.

The marriage has been considered as the divine union that involves recruitment not only for the purpose of biological reproduction but also for the maintenance of an orderly replacement of family culture. According to Dharamshastra the cardinal function of the marriage is to perpetuate one's Kula (lineage) so as to fulfill the social obligation (Dharma). It also couches in terms of psycho-sexual gratification (Kama) but its objective is to attain higher social objective of Kula through ones progeny (Praja). Marriages are of various types e.g. in Daiva marriage the girl is married to a priest during a sacrifice. The Brahma Marriage had prevailed through gurukul system and in Arsha marriage the daughter is given to the Rishi. The Prajapatya marriage is one in which the sacred duties are performed jointly by both the bride and the groom. The Gandharva Marriage is a type of marriage that simply involves exchange of garlands. In Asura Marriage, an unsuitable bridegroom approaches the family of the bride and offers a high price to get married to the girl. In Rakshasa marriage the groom fights battles with the bride's family, overcomes them and takes her away from her family and then forces her to marry. The Paishacha marriage is usually considered as the most inferior type of marriage. There are also the Pre arranged marriage (usually happens among relatives); consummated marriage or celibated marriage among young; and engagement by commissioning, monogamy types of the marriage (the person marrying only to one wife) and polygamous types of marriage. The Polygamy is of two type: Polyandry where one woman has two or more husband and Polygyny where one man has two or more wife. Polyandry is also of two types named as fraternal (sharing of wife by real brothers) as well as non-fraternal (sharing of wife by close relatives). The Dhara Pratha (forceful marriage) is still prevalent in hilly areas of Himachal Pradesh such as Rohru, Chirgaon (Chohara Block), Jubbal, Chopal, and Rampur tehsils of Shimla district. Beside this type of marriage the engaged marriage is prevalent wherein family and Bishtu (closest relatives/neighbourer) play important role in selecting the groom for bride.

The first evidence of attitudes towards sex comes from the ancient texts of *Hinduism, Buddhism and Jainism,* the first of which are perhaps the oldest surviving literature in the world. These most ancient texts, the Vedas, reveal moral perspectives on sexuality, marriage and fertility prayers. The epics of ancient India, the Ramayana and Mahabharata, which may have been first composed as early as 1400 BCE, had a huge effect on the culture of *Asia,* influencing later Chinese, *Japanese, Tibetan and South East Asian* culture. The first marriage in literature was witnessed by'Adam and Eve those according to Zoug did not know about intercourse and private parts. It was first established when the almighty created Adam and Hawwa. In India Jeevan saathi is received through fast, *tap, jap, pooja -path, yajya,* and *havan* etc. There are also trivarga such as D*harma,* and A*rtha, and Kama* concept to be involved in it. In Pativrata dharmathe husband is considered as parmeshwar and wife is called as Dharma *Patni, Dharmini, Ardhangni,* and *Griha Lakshmi.* The *Moksha* can be attained through *trivarga* such as *Dharma, Artha and Kaama.* The Hindu marriage is not a contract like western culture rather is a collision of the two souls. Various practices are performed for marriage in India. They are known as matching Kundli, Muhurat, Kanyadan by father; Apsi Milan, Vivahom, Panigrahana and Agniparinayan, Ahuti; Saptapadi (seven fere) where one is expected for food, power, wealth, santan, rule regulation and friendship.

The first evidence of attitudes towards sex comes from the ancient texts of *Hinduism, Buddhism and Jainism,* the first of which are perhaps the oldest surviving literature in the world. These most ancient texts, the Vedas, reveal moral perspectives on sexuality, marriage and fertility prayers. The epics of ancient India, the Ramayana and Mahabharata, which may have been first composed as early as 1400 BCE, had a huge effect on the culture of *Asia,* influencing later Chinese, *Japanese, Tibetan and South East Asian* culture. As Indian civilization further developed over the 1500 years after the births of

Buddha and Mahavira, and the writing of the Upanishads around 500 BCE, further historical evidence, art, and literature shows that ancient Indian society was perhaps as sexually tolerant as many modern European and East Asian countries. It was somewhere between the 1st and 6th centuries that the Kama Sutra, originally known as *Vatsyayana Kamasutram* ('Vatsyayana's Aphorisms on Love'), was written. This philosophical work on *kama shastra,* or 'love science', was intended as both an exploration of human desire, including seduction and infidelity, and a technical guide to pleasing a sexual partner within a marriage. The *Tantric* school of Indic/ Hindu philosophy gave an idea that sex is a basic and powerful desire experienced by all humans, could be utilized as a way of achieving enlightenment.Nudity in art was considered acceptable, as shown by the paintings at Ajanta and the sculptures of the time. It is likely that as in most countries with tropical climates, Indians from some regions did not need to wear clothes, and other than for fashion, there was no practical need to cover the upper half of the body. This is supported by historical evidence, which shows that men and women in many parts of ancient India mostly dressed only the lower half of their bodies. Beside this an evidence of homosexuality was found by Luth Alaihisalam about 5000 to 6000 yrs ago. In the present scenario it is practiced in western country, American, Asian, and African country those have gained legal status in Holland, Netherlands, and Canada.

In the present manuscript a pioneer attempt has been made to explore the psychological wellbeing among the polyandrous tribal people of Himachal Pradesh. The study reflects the life-style, custom, tradition and culture of tribal people. Special emphasis has been given to explore the psychological wellbeing of polyandrous people who share one wife with more than two brothers. The author is immensely thankful to Sh. Sudhir Choudhary for giving best shape to the manuscript. I hope this preliminary work will be helpful to the students, researchers and teachers pursuing research in the areas.

Dr. Roshan Lal Zinta
Chairman, Psychology Department
Himachal Pradesh University,
Summer Hill, Shimla-171005(india)
E mail: zinta.roshan@gmail.com

CONTENTS

INTRODUCTION

Life is a love affair, it is poetry, and it is music. Don't ask ugly questions like, what is the purpose? Because the moment you ask it, you disconnect yourself from life. Life can not be bridged with philosophical questions. Philosophy has to be put aside, says Osho, the Zen Master

(Hindustan Times, Sept. 26, 2010)

1.1 INTERPERSONAL RELATIONSHIP

The life is a stage with one entrance but with many exits (Durant, 1964). It is truc that biologically the life is premeditated primarily for survival. The brain play pivotal role in life of all the creatures ranging from alligators to human by exerting its influence on body. Personal relationships are loving and supportive with a promising future that leads to romance, commitment, harmony and rhythm in future life. There is a love and joy in personal relationship found in every breath in successful life. One spends time with old friends, family and loved ones to renew bonds and rejuvenate relationships when somebody lost the sharing feeling and ideas. An underlying factor in a very happy person is that they tend to have friendly relationships in their life. According to Wyse (2010) a good friend is a connection to life – a tie to the past, a road to the future, and the key to sanity in totally insane world. In these personal relationships such as awareness, choice, love, kindness, altruism and joy, frontal cortex play important role (Samadarshini, 2010) for maintaining psychological wellbeing. According to Karl Marx love between man and woman is the immediate relationship of human being to the Supreme Being. Therefore, the love is sublime or divine and the beauty of soul.

But in contemporary scenario the meaning of the love has transformed. Now the love has converted into profession as well as sex tourism wherein the traffickers, escorter services and touter are highly involved (Dhawan & Ghosh, 2010) in this business and even do no hesitate to ostracize, pimple or barter the disadvantaged females and girls. By Sam, Dam, Dand or Veda policies they dragged or trapped the women in such an appalling path. For these poor female those mostly belongs to scheduled caste and scheduled tribe categories relinquishing of hunger is bigger problem (Aiyar, 2010) than to protecting their izzat or body. On the one side India is touching the sky by making technological development on the other hand the deprived segments are still soaring even for two comes nutritious time food. Although the government is acting against hunger yet the efforts are not sufficient to curb inflation, of communities so as to extinguish hunger. The poor are becoming poorer and rich as richest. In this direction there is a need of favorable policies by the Govt./non-Government well as certain organization, in general and women's institutions and organization in particular those should come forth for mitigating the war (Kishwar, 2010) of evil and instill the love and affection feeling among the population.

A perfect love therefore produces perfect citizen with balanced personality. But the love before marriage is just infatuation, attraction of the face; it becomes deeper and real after marriage through concordance and confidence. The marriage is the strong basis of a family structure and turning point in ones personal and social life (Kour, 2010). Little girls in the United States hear this all the time, from their mothers to teachers that become a "Sesame Street" characters. Almost everywhere they go, they are encouraged to believe that girls can be just as smart, athletic, and successful as boys but in our country the picture is spunky and full of checkout. The girls or women in our country are conditioned from the day they born by constantly reminding of the things that they are not allowed to do this and that. In such a situation one family play a pivotal role for their development that perhaps requires microanalysis and humanitarian (Bandura, 1977, Zinta, 2008, 2009, 2010) approach to record the change in their behaviour and cognitive processes. One should showed form on Gandhi's success who emphasis for lay convincing India that societies here has became civilized only in men imbibe some of positive famine qualities, such as nurturing and revulsion towards violence, that in turn can make the country on more civilized.

But the love is a statement – it is an action in any respect. It is believed that the men are born from Mars and women from Venus. Therefore, the men are more into hardware stuff like gadgets, tools, computers etc and women are more into social relationships. Love is neither disapproved by religion, nor prohibited by the Law; for every heart is in God's hands." The souls are scattered in the air when they meet and one experience or feel love. Islamic theory of love is based on the principle of Al hubbil Hudri which states that the love is humble and not lustful that perhaps depends on the degree of their lover. Ibn Hazm says that love is natural, but can Allah test us with this? Love is Halal, for every heart is in Allah's hands. It is just happens or just natural. The noble love comprises of passions of heart with righteousness and piety. More appropriately, it is the conjunction between scattered parts of souls that have met in universe. Love is based on assimilations and similarity in characteristics for which physical attraction is not so very important, but liking for someone involves feeling of great or intense affection without any restriction.

Ishq is the deeper love that involves use of intelligence, becomes obsessive and commits haraam deeds whereas the love is a human sickness (Psychological), natural, spiritually imbibe natural instinct full of sacrifice. Muhammad Ibn Dawood – Al Zooah speak about the 100 qualities of Bin Udrah on how he exercised love in his life. Arabs say that the eyes are the white gate into the heart that perhaps involves soul. The three pillars attribute of the beloved one are feelings the hand of love, intention and inclination. The ruling of love in Islam is not in one's hand, but on Allah. The first love in Islam was recognized by Prophet SWS for Ayesha's nurturing love. How to keep love alive, think love is action, practice it on a daily harms love, passions and love, decency or indecency? O' Mankind, be conscious of your duty to your Lord, who created you from a single soul, created of like nature his mate and from the two created and spread many men and women, and be mindful of your duty to Allah whose name you appeal to one another and to (the ties of) the womb. It has imbibed all the quarries in human melding forming interpersonal relationship.

Interpersonal relationships include kinship and family relations in which people become associated by genetics or consanguinity. Relationships can also be established by marriage, such as husband, wife, father-in-law, mother-in-law, uncle by marriage, or aunt by marriage. They may be formal long-term relationships recognized by the law and formalized through public ceremony, such as marriage or civil union. They may also be informal long-term relationships such as loving relationships or romantic relationships with or without living together. In these cases the "other person" is often called lover, boyfriend, or girlfriend, as distinct from just a male or female friend, or "significant other". If the partners live together, the relationship may resemble marriage, with the parties possibly even called husband and wife. A common law can regard such couples as actual marriages after a period of time. Long-term relationships in other countries can become known as common-law marriages, although they may have no special status in law. But in all types of relationship the family play important role, for their long term diversity that is perhaps directly associated with their psychological wellbeing.

1.2 LIFE IN FAMILY INSTITUTION

The status of family in India in particular and in Islam in general play pivotal role for nourishing in intimate relationship as well as in survival. A twice-born man's life is an eternal chain of religious rites and sacraments, beginning from the cradle and ending in the grave. A child is born in the family, and with it starts a series of holy rites and sacraments to admit the new member to the old family circle. The holy rites are performed in order to "sanctify the body and purify (from sin) in this life and after death" (Mukerjee, 1976, p. 65). The family system is divinely the inspired institution. Generally a family is a social unit where the member lives together. It involves an association of people who share common beliefs or activities. An interpersonal relationship is an association between two or more people that may range from fleeting to enduring in the family. It may be based on limerick, love and liking, regular business interactions, or some other type of social commitment.

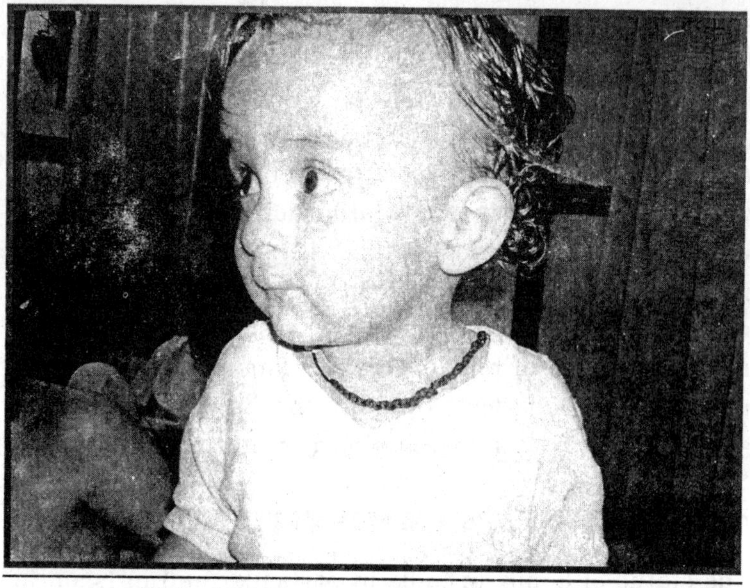

Fig. 1.1.1 (a): Birth Rites of Kinnauri Child

The relationship in family is regulated by law custom, and mutual agreement that further form the basis of social groups and society as a whole. A relationship is normally viewed as a connection between two individuals, such as a romantic or intimate relationship, or a parent-child relationship. Individuals can also have relationships with groups of people, such as the relation between a pastor and his congregation, an uncle and a family, or a mayor and a town. The intimate relationships are, however, only a small subset of interpersonal relationships. The model mentioned below hope to improve our understanding of the ways in which family resources influence family management, beliefs, and caring behaviours, including emotional climate and child-care quality within the home, and through which family management, beliefs, and caring behaviours influence the development of the children though interpersonal relationship. Because child's development is measured multidimensionality, and the model integrates several family components that potentially influence each dimension of child development. Beside child the well-being of other family members is similarly determined by structural equation modeling with latent variables (Bollen 1989; Joreskog & Sorbom 1989). People in a relationship tend to influence each other, share their thoughts and feelings, and engage in activities together.

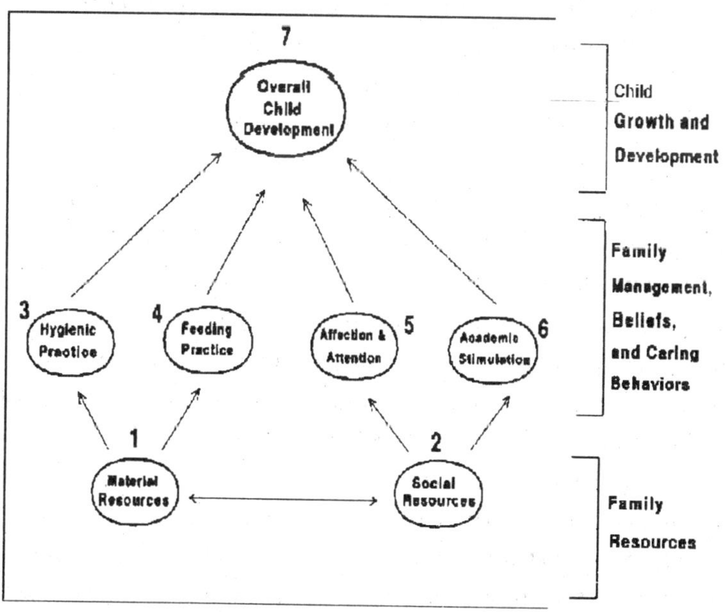

Fig. 1.1.2 (b): Family Functioning Model

The study Myers (1992) distinguished between the definitions of child growth and child development. He defined growth as "a change in size," whereas development was defined as "changes in complexity and function PGF." For practical purposes, we use the term "child development" to refer to both child growth (as measured by anthropometric indicators) and child mental development (characterized by the mental development index or IQ, the physical development index for younger children, and the social development index for older children). The terms are separated for measurement and analysis purposes, however. The general family functioning model has various levels. The first level of this model, deal with "family resources, it has includes both material (no. 1) and social (no. 2) . Many factors at this level of model are measures of socio-economic status, such as housing, food budget, literacy, educational level, and media use. These factors feed into the next level of the model, "Family Management, Beliefs, and Caring Behaviors." These encompass measures of hygiene (no. 3) and feeding practices (no. 4), parents' care giving behaviour (affection and attention, no. 5), and academic stimulation (no. 6). These factors, in turn, determine the child's growth and development (no. 7).

Although humans are fundamentally social creatures, the interpersonal relationships are not always healthy. Examples of unhealthy relationships include abusive relationships and co-dependence. Most scholarly work on the relationships focuses on the romantic partners in pairs or dyads. The interpersonal relationships include friendships, such as relationships involving individuals providing relational care to marginalized persons. These relationships usually involve some level of interdependence.

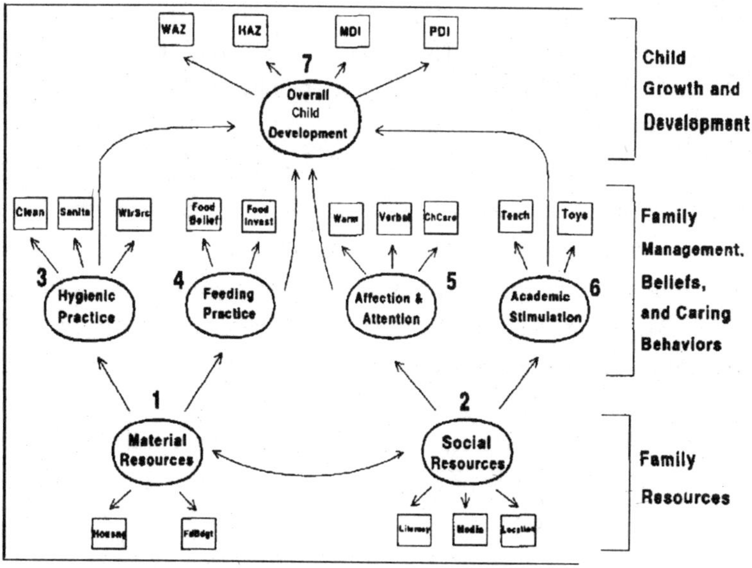

1.1.3 (c): Detailed Conceptual Model, for overall Child Development.

The figure presented above is a more detailed version of the model. Additional factors are community endowment and support from within the family. These two factors are conceptually related to socio-cultural support. That may influences family management and caring behaviors. The socio-cultural support includes support from the family and other kin (the father's involvement in child care and the mother's satisfaction from support received from family and relatives) and community support/endowment (the number of health facilities and community health indicators). Literature indicates that fathers also are involved in child-care activities, especially after the child has begun to walk; therefore, it is culturally appropriate to test this aspect in all the setting with samples of older children. Modernizing lifestyle is expected to influence parental care behaviors. This includes "modern" child-rearing behaviors, such as parents eating together with the child, positive affection rather than negative disciplinary behaviour, and a delay in requiring the child to perform housework. In the polygamous culture of Nigeria, it includes the level of the father's involvement in the household.

In the family the child's characteristics are hypothesized to influence parental care behaviors. Professionals in child development increasingly acknowledge the influence of the context in which the individual child grows. Bradley (1989) emphasized the importance of understanding parents' responses to other family factors, such as income and social status, although there is little research in this area by child development specialists. Family resources alone are not enough to promote positive child development, however, much depends on how the family translates the available resources into positive child-caring behaviors. Many researchers believe that interventions or actions to improve the caring behaviours within the family can be implemented with little or no improvement in family income (Engle 1992).

Parental education, especially maternal education, is consistently related to improved child nutritional status and lower infant mortality (Heller and Drake 1979; Behrman and Wolfe 1984), better child feeding and health practices (Zeitlin et al. 1988), a higher level of child stimulation (Zeitlin, Ghassemi, and Mansour 1990), greater use of health care services (Grant 1984), lower fertility, and more child-centered care behaviours (Ware 1984). Some researchers have argued that the positive association between mother's schooling and child health is not due primarily to the effect of schooling per se; instead, schooling serves largely as a proxy for unobserved characteristics related to her childhood background. A study in Nicaragua that controlled for the heterogeneity in unobserved maternal endowment through the shared childhood experience of adult sisters found that the impact of the mother's education declines substantially. Behrman and Wolfe (1987) also found that maternal schooling effects evaporate when the maternal endowment (abilities, habits, and health status related to childhood family background) is controlled for, thus raising doubts about standard estimates of the impact of maternal schooling on child welfare.

1.3 MARITAL SACRAMENT IN HINDU INSTITUTIONS

One of life's greatest joys is a successful marriage. Few other relationships bring as much satisfaction and fulfillment as a close physical and emotional partnership with one's spouse. While it is easy to get married, it is much harder to remain happy ever after. But with patience, commitment and dash of humour it can be forfeited. Indians regard the institution of marriage as a sacrament. For Hindus, it is the commitment that lasts forever and is the strongest social bond between a man and a woman. The contemporary way of marriages and weddings in India is very different from what it was years ago. But the trend was entirely different and conservative in ancient times. Marriages have always been considered as a divine union in India. The institution of marriage here can be seen from two points of view: the social or objective and the individual or subjective. The objective aspect relates to the values and norms underlying the institutions. According to Dharamshastra the cardinal function of the marriage is to perpetuate one's Kula (lineage) so as to fulfill the social obligation (Dharma). It also couches in terms of psycho-sexual gratification (Kama) but its objective is attain higher social objective *i.e.* to run ones Kula through ones progeny (Praja). Marriage, thus involves recruitment not only for the purpose of biological reproduction but also for the maintenance of an orderly replacement of family culture (Farber, 1964).

Beside this the subjective indicator focused with one caste status so as to maintain caste solidarity and identity, social differentiation and caste system. At the subjective level function of endogamy are similarly crucial. Here marriage arrangement deals with caste background, language, food habits, life style, cultural identities, socialization and family culture (Goode, 1967).Hence the subjective goal of marriage should be in terms of romantic love, conjugal bliss and similar aspects. The parental role in the choice of partners and preferential mating, and the negative emphasis on dating, courtship and premarital love, tend to support the continued importance of marriage in relation to the creation and maintenance of condition effective for the perpetuation of caste organization (Ramu, 1977). Marriage thus in most of India in the past and now generally has never been considered only as an affair between two people (Ross, 1960) rather an agreement between two families.

Despite this, some youth in the contemporary scenarios are applying fatalistic approach never adhere customs of social institution toward marriage and do not view premarital romance as necessary and emphasizes upon the concept that the marriage are made in heaven. According to Hindu view of life the marriage is successful which transform a chance mate into a life companion (Radhakrishnan, 1942). There are many opportunities for boys and girls of marriageable age to come into contact with each other. Their examples are schools, college, offices and clubs where they maintain physical proximity, cycle of dating, courtship and engagement. But the parents or family comes in between their romantic relationship on the concept of jati, and gotra over shatter their dream that in this contemporary scenario do not fit and the advanced youth perhaps do not bother such practice, but still this system exert enormous pressure on their well-being its example in India is Haryana where Khap Panchayatas has given serious punishment to the lower based on caste. Allah SWT principles also called marriage a (Methaakan Galidah) "strong and dangerous covenant" – a serious social contract. Marriage in Islam is the only legal way to express love to your spouse.

In the Vedas too it has been mentioned that a person after the completion of his student life should enter the second stage of life that is the life of a householder. These days, Hindu matrimony are celebrated very lavishly by family and friends. Not only this, parents very meticulously search for the most prospective match for their children who are in a marriageable age. The contemporary way of marriages and weddings in India is very different from what it was years ago. But the trend was entirely different and conservative in ancient times. The Brahma Marriage prevailed gurukul system, where the boy used to go to live with his guru to attain worldly knowledge and expertise. This stage is called Brahmacharya or student hood. The boy would be eligible to get married only once he completes his studies i.e. his student. In Daiva marriage the girl is married to a priest during a sacrifice. The parents of the daughter wait for a suitable man till sometime and when they do not find anyone, they start looking for a groom in a place where the sacrifice is being conducted. At this place, the girl is groomed with ornaments and is married to a priest. As per the Shastras, the Daiva marriage is inferior to Brahma marriage because it is degrading for the woman to look for a groom like that. In Arsha marriage the daughter is given to the Rishi (sage) and the bride in exchange for two cows, which they receive from the groom"s side. This type of marriage use to take place because it was difficult for the parents of the bride to afford the expense of their daughter's marriage at the right time as per the Brahma rites. If that didn't happen, the girl was made to marry an old sage. Further, the Prajapatya marriage is one in which the sacred duties are performed jointly by both the bride and the groom. For this type of marriage, Swami Dayanand Saraswati, the founder of Arya Samaj, said that the father gives his daughter to the bridegroom, by addressing the couple with the mantras, "*May both of you perform together your dharma*". This kind of marriage is quite similar to Brahma vivah, which can be described as the origin of Prajapatya vivah.

The Gandharva Marriage or Gandharva Vivaha or the marriage of the celestials is a type of marriage that simply involves exchange of garlands after which the marriage is confirmed. The Gandharva marriage is equivalent to eloping in the contemporary world. It is considered that couples whose union is not blessed by families seek refuge in this kind of custom. In Asura Marriage, an unsuitable bridegroom approaches the family of the bride and offers a high price to get married to the girl. It could be considered as a bribe for marrying the girl whom the boy desires even if he does not deserve for her at all. In Rakshasa marriage the groom fights battles with the bride's family, overcomes them and takes her away from her family and then forces her to marry. The Paishacha marriage and is usually considered as the most inferior type of marriage. In this marriage, no one considers the girl's wishes, whether she wants to marry or not instead. Temptation and hectic life style put marriages under stress. A successful marriage is one where the couple has sex regularly, though not with each other (HT, September 19, 2010).

The marriage therefore is a legal union between one man and one woman as a set of cultural rules for bringing men and women as well as the family together. The marriage is the cooperation and understanding of the rights of each other and their obligations. It is an ancient practice as taken a life's term companion for sexual partner or a civil contract between a man and women for living as husband and wife together. It is also considered as a legally recognized and or socially approved bond between two individuals that carries certain rights and responsibilities for sexual activities. More appropriately the marriage is a contract between a man and a woman, which allows both to enjoy the person of one another. The first marriage in literature was witnessed by'Adam and Eve. According to Zoug they did not have intercourse in Jannah, because they did not know about their private parts. It was first established when the almighty created Adam and Hawwa. Therefore the oldest family to the mankind is the marriage of Adam and Hawwa .

The marriage has it remained as the primary institution in India during Vedic period and is considered as a ritual or *sanskar,* important for shudhi, saphai, reform the mind and body, physical, social and spiritual as well as religious affairs. It is the bond between ecstatic and domestic life. It is considered as the ritual that involves mate selection through Kundli, Muhurat, Kanyadan by father; Mutual glance, Vivah hom, Panigrahana and Agniparinayan or Ahuti; Saptapadi (seven fere) where one is expected for food, power, wealth, santan, rule regulation and friendship. It is also considered as the ritual where one enters to grihastha ashram, performed ceremonies and fasts like Karvachoth. The marriage is the sacred bond that is settled in heaven but celebrated on earth. Ross(1960:282) meaningfully pointed out that, "formerly the rarely marriage of women and their incorporation into their husband's household at a very impressionable age meant that even the minutest detail of family behaviour could be transmitted though them to the next generation.

The problem of a satisfactory definition of marriage has yet vexed anthologists for decades and has been raised, but not solved, several times in recent years. Over times it has become clear that the "Marriage is a union between a man and a woman such that children born to the women are recognized as legitimate offspring of both parents" (Royal Anthropological institute, 1951). The marriage is a union between one man and one woman, and that it establish the legitimacy of children (Evans-Pritchard, 1951). By implication, then, whatever may be an individual's efforts in securing a bride the feeling, in general, is that a man gets the bride ordained for him. According to the Hindu view, the marriage is successful which transforms a chance mate into a life companion (Radha-Krishnan, 1942:85). The writing on marriage in India falls into two broad types. On the one hand we have example of descriptive anthropology in which marriage practices (monogamy, polygyny, polyandry, levirate, arrange marriage', etc.) are seen as significant cultural traits to be interpreted and explained in reference to other aspects of the social structure and physical environment. Hindu ideals encourage pre-puberty marriage for girls (Kapadia, 1966). Of the several explanation for this are: (1) pre-puberty marriage helps ensure the girl's virginity, for if a girl is given in marriage when she is too young to experiment with sex, it is assumed that her virginity is preserved and (2) when girls does marriage at an early age, it means that they move directly from the tutelage of their parents to the tutelage of their parents-in-law.

In India Jeevan saathi is received through fast, *tap, jap, pooja path, yajya,* and *havan* etc. There are also trivarga such as *dharma,* and *artha, and kama* concept to be involved in it. In Pativrata dharmathe husband is considered as parmeshwar and wife is called as Dharma *Patni, Dharmini, Ardhangni,* and *Griha Lakshmi.* The *moksha* can be attained through *trivarga* such as *Dharma, Artha and Kaama.* The Hindu marriage is not a contract like western culture rather is a collision of the two souls. In Vedic period, Sutra, *Epic period, Age of Legislators, Puranas commentators, Digest writers* and *Mahakavya,* it has its explanation that the age for marriage should be between 18 to 21 years. Get rid of the "ideal" that somebody has to be married and maintain that facade while screwing for variety and many of the negative aspect vanish. Hindu tradition stipulates that the important objects of marriage are religious duty (dharma), progeny *(praja)* and conjugal love *rati*), and Hindu traditions consider marriage more as an instrument to be used in pursuing the higher goals of life, define in terms of religious responsibilities, than as an arrangement for personal comforts and conjugal pleasure. In fact, it considers the latter to be related to marriage only to the extent that they aid in the perpetuation of one's family culture. In short, marriage is a social obligation rather than a provision for an individual's private pleasure (Prabhu, 1949: Kapadia, 1966)

The sexual innovators, over the past 60 or 70 years, have argued for a third alternative—a combination of permanence with permissiveness that is, permanent adherence to the marriage, for the sake of child-rearing and social stability, combined with freedom for each partner to have additional emotional and physical relationships outside the marriage. First importance is the continuity of the family line. The sexual pleasure and personal comfort have been ranked second and emotional gratification third. Although value related to conjugal love and other emotional concerns are important. Yet, all variations upon this theme have proven disruptive to the marriages of most of those who have practiced them, and too threatening to the majority of those who have not to be seriously tried out. Relatively few people, even today, manage to make permissive marriage work at all, let alone work better than exclusive marriage. The marriage was not clear cut in the Christians era those only were following Jews – after separated from Jews, didn't' have anything, except getting married in a Church – have no mahr, no concept of family as well as of religion. The marriage not only with one wife one husband is prevalent in India rather than with more is practiced here.

1.4 POLYGAMOUS LIFE STYLE

The human beings by nature are polyandrous. Its example can be taken from the Mahabharata (Book 1, Chapter, 122) wherein it is stated that women in primitive times were free and not bound to be faithful to their husbands. They could have intercourse with any man they chose, even though they were bound by ties of marriage (Parmar, 1975). India played a significant role in the history of sex, from writing the first literature that treated sexual intercourse as a science, to in modern times being the origin of the philosophical focus of new-age groups' attitudes on sex. It may be argued that India pioneered the use of sexual education through art and literature. As in all societies, there was a difference in sexual practices in India between common people and powerful rulers, with people in power often indulging in hedonistic lifestyles that were not representative of common moral attitudes. Indian culture can be considered amongst the most ancient, with the ancient Indus Valley civilization being contemporary to ancient Egypt and Sumer, spreading across modern India and Pakistan at its peak, 4000 years ago. During this period, not much is known about social attitudes toward sex. One thing that has been observed about sexuality in the Indus Valley civilization is the practice of fertility rituals. Early philosophy and theology related to sexuality may have developed during this time.

The first evidence of attitudes towards sex comes from the ancient texts of *Hinduism, Buddhism and Jainism,* the first of which are perhaps the oldest surviving literature in the world. These most ancient texts, the Vedas, reveal moral perspectives on sexuality, marriage and fertility prayers. The epics of ancient India, the Ramayana and Mahabharata, which may have been first composed as early as 1400 BCE, had a huge effect on the culture of *Asia,* influencing later Chinese, *Japanese, Tibetan and South East Asian* culture. These texts support the view that in ancient India, sex was considered a mutual duty between a married couple, where husband and wife pleasured each other equally, but where sex was considered a private affair, at least by followers of the aforementioned Indian religions.

As Indian civilization further developed over the 1500 years after the births of *Buddha and Mahavira,* and the writing of the Upanishads around 500 BCE, further historical evidence, art, and literature shows that ancient Indian society was perhaps as sexually tolerant as many modern European and East Asian countries. It was somewhere between the 1st and 6th centuries that the Kama Sutra, originally known as *Vatsyayana Kamasutram* ('Vatsyayana's Aphorisms on Love'), was written. This philosophical work on *kama shastra,* or 'love science', was intended as both an exploration of human desire, including seduction and infidelity, and a technical guide to pleasing a sexual partner within a marriage. This is not the only example of such a work in ancient India, but is the most widely known in modern times. It is probably during this period that the text spread to ancient China, along with Buddhist scriptures, where Chinese versions were written. The *Tantric* school of Indic/Hindu philosophy formed at some point in this period, and part of the philosophical system was the idea that sex, as a basic and powerful desire experienced by all humans, could be utilized as a way of achieving enlightenment. Some ardent devotees of this system for example might deliberately break sexual taboos that were ridiculed, such as extramarital sex, to master human nature and achieve greater understanding of the universe, their soul. The Tantric tradition spread throughout Asia as far as Japan.

It seems that polygamy was allowed during ancient times. In practice, this seems to have only been practiced by rulers, with common people maintaining a monogamous marriage. It is common in many cultures for a ruling class to practice polygamy as a way of preserving dynastic succession. Nudity in art was considered acceptable, as shown by the paintings at Ajanta and the sculptures of the time. It is likely that as in most countries with tropical climates, Indians from some regions did not need to wear clothes, and other than for fashion, there was no practical need to cover the upper half of the body. This is supported by historical evidence, which shows that men and women in many parts of ancient India mostly dressed only the lower half of their bodies. Whilst this has changed in modern times, it is likely that taboo against nudity was not presented in many Asian, African and South American civilizations and the taboo in Europe is a matter of climatic necessity.

The marriages across the world in general and in India in particular are of various types. It involves Pre arranged marriage (usually happens among relatives); consummated marriage or celibated marriage among young; and engagement by commissioning. In monogamy types of the marriage, the person marrying only one wife. In polygamy, there is a practice of having more than one spouse at one time.

Fig: 1.1.4 (d): Polyandrous life style in district Kinnaur

Polygamy is of two type: Polyandry where one woman has two or more husband and Polygyny where one man has two or more wife. Polyandry is of two types named as fraternal as well as non-fraternal. In fraternal polyandry the wife is shared by the real brother where as in non-fraternal one it can be shared even by other close relative beside the family brothers. It is also clear that even in societies that institutionalized some form of polygyny, monogamy was nonetheless frequent, although, for men at least, this typically meant making the best of a bad situation. Certainly there is no evidence, either from biology, primatology, or anthropology, that monogamy is somehow 'natural' or 'normal' for human beings.

Although today, polygamy (having more than one spouse) is not legal across the world including United States and India, there was a time when many religions embraced such a concept. In fact, it was considered a blessing for Jews to have more than one wife at one time, and Mormons, too, adapted this concept to perpetuate the survival of their own people. Some cultures, especially in Africa, today, even sanction the marriage of one man to more than one woman, and vice-versa. Often these individuals share households, as well as duties and responsibilities, thereby creating an appropriate division of labor. Polygamy may appear to be attractive, especially for those who complain of sexual dissatisfaction within their own marriage, but underneath the veil of sex comes many other responsibilities as well. The concept of polygamy is the natural extensions of family and community which, in an ideal setting, can only contribute to the growth, support, and learning for all members involved.

The heterosexual genital love, which has remained exempt from outlawry, is itself restricted by further limitations, in the shape of insistence upon legitimacy and monogamy. Present-day civilization makes it plain that it will only permit sexual relationships on the basis of a solitary, indissoluble bond between one man and one woman, and that it does not like sexuality as a source of pleasure in its own right and is only prepared to tolerate it because there is so far no substitute for it as a means of propagating the human race. The people who used polygamy are involved in various sexual activities with many women as well as even men. The homosexuality that refers to the attraction to the opposite sex has also attracted the attention of the researcher since time immemorial as well in the contemporary scenarios. It was first known in the times of Luth Alaihisalam about 5000 to 6000 yrs ago. It was also practiced in ancient China, Native Americans, ancient European times, and Africa etc. as well as accepted in Holland, Netherlands, and Canada. Across the world in general as well India in particular the population of homosexual e.g. lesbian and gays have been considerably increasing. Either they want to get rid of their psychological tension or may be the other socio-psychological and cultural reason or the economic circumstances behind attitude and behaviour.

In nations that allow polygamy, women often feel such practice of to makes them as second class citizens. The women's rights movements in these nations want to make monogamy the only legal form of marriage. The United Nations joined these efforts in 1979 when the General Assembly adopted the Convention on the Elimination of All Forms of discrimination against women. An international bill of rights for women appeared in front of over 180 nations those have agreed to implement Article 16 of the Convention requires nations to give women and men equal rights of marriage. Polygamy is interpreted as inconsistent with Article 16 because it extends the right of multiple spouses to men but not to women. The United Nations has established the Committee on the elimination of discrimination against women *(CEDAW)* to monitor the progress of nations implementing the convention. The United Nations is thus working through the Convention and CEDAW to promote women's equality by making monogamy the only legal form of marriage worldwide. The African Union has recently adopted the *Protocol on the Rights of Women in*

Africa entered into force November 25, 2005. But the protocol does not suggest making polygamous marriage illegal. The article 6 of the protocol states that "monogamy is encouraged as the preferred form of marriage and that the rights of women in marriage and family, including in polygamous marital relationships are promoted and protected.

The 1950 Marriage Law called for sweeping changes in many areas of family life. It forbade any 'arbitrary and compulsory' form of marriage that would be based on the superiority of men and would ignore women's interests. The new democratic marriage system was based on the free choice of couples, monogamy, equal rights for both sexes, and the protection of the lawful interests of women. It abolished the begetting of male offspring as the principal purpose of marriage and weakened kinship ties which reduced the pressure on women to bear many children, especially sons. With arranged marriages prohibited, young women could choose their own marriage partners, share the financial cost of setting up a new household, and have equal status in household and family decision-making. The Government initiated an extensive campaign of marriage-law education, working jointly with the communist party, women's federations, trade unions, the armed forces, schools and other organizations.

But the present-day civilization makes it plain that it will only permit sexual relationships on the basis of a solitary, indissoluble bond between one man and one woman, and that it does not like sexuality as a source of pleasure in its own right and is only prepared to tolerate it because there is so far no substitute for it as a means of propagating the human race. None of the many restrictions that Christianity has placed upon sexual expression has been more highly valued—and more burdensome—than the doctrine that husband and wife must limit themselves sexually to each other from marriage until death. Our pious great-grandfathers may have regarded this as the natural and only moral way of life. It is so rare a pattern in anthropological and historical perspective that one is forced to consider it, if not unnatural, at least idiosyncratic and no more moral than any one of a hundred other alternatives. Currently, monogamy is the only love style considered legitimate by our culture, even though the evidence clearly indicates that humans are not monogamous by nature. Monogamous couples are completely dependent on each other for affection and sex; and many become dissatisfied due to sexual incompatibilities, differences in level or frequency of sex, boredom with their sexual patterns. The defense of monogamy is as varied and rich as the criticism of monogamy. Many authors claim that sexual monogamy promotes security, intimacy, and stability in relationships. Their claim stems from observations of couples exploring open marriage. Although some people have happy and stable open marriages, sexual non-monogamy proves too difficult for most couples to manage and their relationships suffer as a consequence. Many authors criticize lifelong sexual monogamy as unnatural and unrealistic. They contend that humans have never been a sexually monogamous species, and that cultural expectations of sexual monogamy place enormous burdens on individuals to fulfill all the sexual needs of their partners.

1.5 POLYANDRY AND PSYCHOLOGICAL WELLBEING

The Himalayan countries have the custom of polyandry (Parmar, 1975). It is prevented in tribal areas. The word 'Tribe' denotes a group of people living in primitive or barbarous conditions. It is a social group with territorial affiliation, endogamous with no specialization of functions. They have a headman or a chief who controls the activities of that group. There are various rudimentary customs and traditions among the tribal. For example the people living in the area of Himachal Pradesh are polygamous they either practice polyandry or polygyny system. Most of the tribal in district Kinnaur of Himachal state is fraternally polyandrous where sharing of one wife is common among various brothers.

Fig.1.1.5(e): Dress-up style of Dulha and Dulhan in District Kinnaur in Marriage

The tribal have several sub-groups all of them together known as 'Tribal Society'. It is difficult to say whether they are indigenous or not but they are earliest settlers of India. They were living in forests since early times and even now some of the groups follow the same trends and live in forests. India has the largest population of tribal (Adivasi) groups in the world with a population of 8.43 crore which is approximately 8.2% of the total population there are of more than 100 crores. They are spread over the long corridor of Western, Central, and Eastern states of India. The highest concentration of tribal are in the states of Bihar, Jharhand, West-Bengal, Orissa, Chattisgarh and Assam. Beyond the Indian borders they are also found in Bangladesh, Nepal and other places in sizeable numbers. Since they are older settlers and living in forests, they are known as Vanyajati, Vanvasi, Pahari, Adivasi, Anusuchit Jati, Anusuchit Jamati, etc. in Indian languages. Seventy-five of the 698 Scheduled Tribes are identified as Primitive Tribal Groups by considering them as more backward than Scheduled Tribes. They continue to live in a pre-agricultural stage of economy and have very low literacy rates. Their populations are stagnant or even declining. The Kinnaur district has Kinnaura tribes those have their rudimentary system and also are polyandrous in nature. They also follow caste system and low caste people for there are untouchable who are socially discriminated.

Fig:1.1.6 (f): Lada along with Kones in District Kinnaur in a traditional dress of marriage (Janekang)

Polyandry marriage refers to "A relationship established between a woman and two or more other persons, which provides that a child born to the women under circumstances not prohibited by rules of the relationship is accorded full birth-status rights common to normal members of his society or social stratum. So as far as its types are concerned polyandry has been divided into two types depending upon the kind of relation the different husbands have among themselves. The husbands may be brothers or not at all related. Thus this relational affinity divides polyandry into two: 1. Fraternal Polyandry in which husbands are all brothers: 2. Non-Fraternal Polyandry in which husbands are non-related. Fraternal

Polyandry in this the husbands of the woman are all brothers. This type of polyandry is common among the *Khasas* of Dehra Dun. Among other tribes where fraternal polyandry is present in some or the other form, mention may be made of the *Gallongs Mala Madessars, Mavilans*, etc. of Kerala. Rivers opines that the *Todas* practise fraternal polyandry but it is not much in practice these days. Non-Fraternal Polyandry In non-fraternal polyandry, the husbands of a woman need not be fraternally related to each other. A woman is free to choose partners from among the persons other than her husband's brothers. She successively lives in the apartment of her different husbands and while she is staying with one, the other husbands have no right to enter. Among the Todas of Nilgiri a woman has perfect liberty to choose any individual as her mate. The woman, having several husbands, makes the arrangement in such a way that she spends the first month with, say, the first husband, the second month with the second, the third, and so on, or, according to as she allots the months. The Nayars are considered to be a specific case for the non-fraternal polyandry. Aiyappa is of the view (in our personal discussions) that this type of polyandry or polyandry as such has now become a thing of the past. The *Kotas* also practice this kind of polyandry. Against it to some extent are found tribes like the *Karvazhies, Pulayas, Muthuvans and Mannans* in Kerala. What lead to the origin and development of polyandry? Prince Peter in his work on polyandry (The Study of Polyandry), categorized the various causes reasons (for polyandry) into five categories namely historical, demographic, sociological, economic, and personal. But the most vital and important cause deals with ecology/ environment has not been included as a distinct important cause.

Historical View Majumdar writing on Jaunsaries people inhabiting Jaunsar-Bawar of Dehradun district, U.P. remarks that polyandry in Jaunsar-Bawar has been reinforced the mythology of the Pandavas from whom the Khasas trace their origin. On the other hand, Prince Peter, (1963) writes with regard to the polyandry of the Mahabharat in India, Vysasa says that the custom of taking more than one husband has existed in the country since time immemorial. The Vedas, the Sutras and the Smritis, the Mahabharata and the Ramayana, the Buddhists texts as well as the Jains, Kautilya and the Kamasutra all attest its existence in early India. And the evidences of literature are reinforced by the later iconography and art and by the presence of polyandrous pockets from Tibet in the north right down to Ceylon in the South. The equivocally of the prevalence of polyandry in India in the past, especially when Draupadi got married, is very clear from earlier writings. Rigveda certainly permits polygamy though monogamy may have been the rule probably, polygamy, though allowed, was practically confined to the Rajanya class, polyandry is not referred anywhere in the Rigveda.

There are approximately 300 castes in India deals that with the Hindus traditional and have picked up the concept from religious books such as Veda, Dharamshastra, Brihdaranyak Upanishad and Samrityyan e.g. Manusamriti. The Rig-Veda is the older book in the world where there are description of the Caste at (10-99-12) as *"Purushsukta mein Brahamnoasya Mukh Padhbhyan Shudroajayat"* i.e. from the Mounth of Brahma Brahamna originated; from Arm Kshatriya originated, from Jangh, vaishya and from Per(leg) Shudra originated or give rise to birth. It is believed that in Mahabharata's Shantiparva, SriKrishna gave birth to 100 Brahaman, 100 Kshitriya , 100 Vaishya and 100 Shudra and on this basis the world created and caste system evolved. But Gita all four varnas description is based on their *karma and gunas*. The origin of caste specifically deals with food, marriage violation and social living. Evolutionary theory of Denzil Ebbetson states that the caste originated though religious guilds. Hutton highlighted role of racial thought in its primitive theory of culture. He thought caste system as wonderful social institution where one must discover it in India only that specifically deals with food obligation. According to them Manna is a secret power found in people that goes by touching to other. Hence in some extent the origin of caste system and untouchability originated that is still prevalent in all the states of India, Himachal Pradesh in general and District Kinnaur and tribal areas in particular.

Manu considered *Pratilom Vivah* and Varna *sankarta* as an example of caste origin. Racial theory of Hebert Risley, Ghuriye, Majumdar, Radha Krishan, S. C., Roy, N. K. Dube, MacIver, Maxweber, Kroeber considered origin of castes from racial difference. According to them when Aryan People came from Persia to India. Their society was divided into four parts and our country adopted their system and resulted caste system. According to Dubois's political Brahminical theory, it is the clever plan of Brahaman to form caste system to make their supremacy and higher social status. Occupational theory of Nesfield states that it the occupation is the main factor behind caste origination. Nesfield divided society in three parts. Purohit, Shasak, and Poonjipati Varga those form Dwij and finally guild. A theory formulated by Hokarts, and Senart de also with religious activities as the main factor behind castes system. For example main function of Brahaman to say mantra, Mali as to pluck flower and Kumhar as to make pot.

In Kinnaur Daloch, is known as dab-dab fiyo fayo is the famous name for forceful marriage and the Haar is its one form where other unidentified male forcefully picked up any married women and forcefully maintain relationship. The angry people are given izzat with money and Bakra known as Handa (Gold items) and Bakra (goat) after forceful marriage for maintaining cordial relationship in later life. The kinnaur district also follow various marital customs that includes Janekang (Arranged Marriage); Damchalshish/ Damtangshish/ Bennang Hachis/ Jushish (Love marriage) and Darosh/ Dab-Dab/ Nyamsha/ Nyamsha- Lemo / Ashish / Huchis (Forceful marriage) and Her (Haar or Enticing others wife). Although there do exist different types of marriage in Kinnaur yet it is all linked with the promotion of psychological wellbeing.

In recent years, psychological well being (SWB) is the focus of intense research attention (Diener & Diener, 1995). With increasing recognition of the different streams of inquiry guiding the broad domain of psychological well being, Ryan and Deci integrated two broad traditions: one dealing with happiness (hedonic wellbeing) and other dealing with human potential (eudaimonic wellbeing);(Ryan,2008). Psychological well-being resides within the experience of the individual (Campbell et.al. 1976). It is person's evaluative reaction to his or her life—either in terms of life satisfaction (Cognitive evaluations) or affects (ongoing emotional reaction). Diener and Diener (1995), however, held that little attention has been paid to whether the predictors of psychological well-being differ in various cultures. They further added that variables that influence peoples' evaluations of their well being do vary across culture. While working in the area of psychological well-being in India an author has pointed out the role of personality and social support in psychological well-being. Whereas, Diener et al., 1984, held that people who are successful at attending frequent positive affect will be happy. There appears a movement towards holistic view of positive mental health at the community level. In addition, one needs to consider the role of geography, culture and history in mental health. Vaillant (2003) has contrasted six different empirical approaches to mental health. The first model, having long history, conceptualizes mental health as above normal. The second model also has a long history where health is compared with positive psychology approach. The third model takes a developmental stance where mental health is associated with maturity. The fourth model of mental health considers the importance of socio-emotional intelligence. The fifth model can be conceptualized as subjective well being, a mental state that is subjectively experienced as happy, contended and desired. The last or the sixth model conceptualizes mental health as resilience in maintaining homeostasis and better adaptation. Beside this

There are two important paradigm shifts in understandings individual and his relationship with social environment. The first paradigm shift involves the acceptability of the social science of behaviour, and the second paradigm shift relates to changes within the behaviour therapy movement. These two paradigm shifts are, respectively, concerned with the evolution of social behaviour and the role of social factors in strengthening inner resources of human being. It is also believed that the human behaviour, particularly social behaviour, develops through learning, the biological factors are not important . The new paradigm in social behaviour takes into consideration innate and learned factors following evolutionary approach. This is particularly true in the area of mental health, which is defined as the mental, social and physical wellbeing of the person. As far as social psychology is concerned there are various paradigms of psychological changes related to psychological well being. Socio psychological paradigms conceptualize psychological processes in terms of cultural and genetic co-evolution. It relates personal identity to social approach as a result of which, individuals can adopt values and behaviours that do not support their wellness. (Crook & Osmaston,1994). As per Eco-cultural paradigm, social and political contextual variables are incorporated and they are seen in a dynamic interaction with ecological variables that brings biological adaptation(Mishra, Sinha & Berry,1996). Native cognitive model includes the study of group dynamics in a co-evolutionary approach. It considers ecological and psychological parameters and involves assumptions of mind, experience of people, grass root orientation to social problems and cohesiveness among the members of groups (Pirta, 2002-2003).

POLYGAMOUS LIFE STYLE IN TRIBAL AREAS

Man is alive only inasmuch as he is productive, inasmuch as he grasp the world outside of himself in the act of expressing his own specific human power, and of grasping the world with these powers, inasmuch as man is not productive, inasmuch as he is receptive and passive, he is nothing, he is dead

—Karl Marx

2.1 EPITOMIZING PREMIUM LIFE

The earth is only the planet in the universe that is very well known to harbour life on it. Evidence suggests that the life on the earth existed about 3.7 billion years ago. A creation myth is a supernatural story that even describes the beginnings of the "universe" (cosmogony), "earth" as well as the "humanity" or the "life" often as a deliberate act by one or more deities.The origin of life attempt to find a mechanism that explains about the formation of a primordial single cell organism from which all life originated. A Scottish scientist, James Hutton in 1785, the father of "Geology", propounded it by stating that the earth once was a super organism. His idea of living earth later vanishes in the intense reductionism of the nineteenth century. Similarly, the Gaia hypothesis as proposed by scientist James Lovelock in 1960 had explored the idea that the life on earth functions as a single organism that in turn defines and maintains environmental conditions that has prime importance in the survival.

According to Aristotle, all things in the material universe have both matter and form. The form of a living thing is its soul (Greek 'psyche', Latin 'anima') and matter the body. There are three kinds of souls: the 'vegetative soul' , found in plants, causes them to grow, decay and nourishment; the 'animal soul' which causes animals to move and feel; and the rational soul which is the source of consciousness and reasoning that Aristotle believed is found only in human being. Each higher soul has all the attributes of the lower one. In biology, the science of living organisms, "life" is the condition which distinguishes active organisms from inorganic matter, including the capacity for growth, functional activities and the continual changes preceding to death.

The diversity of life on earth today is a result of the dynamic interplay between genetic opportunity, metabolic capability and environmental challenges. The inert components of an ecosystem are the physical and chemical factors those are necessary for life – energy (sunlight or chemical energy), water, temperature, atmosphere, gravity, nutrients, and ultraviolet solar radiation protection. Since the inception of life on the earth, the human beings have been continuously struggling for surviving better in their environment. Due to various socio-psychological and cultural factors, the paces of human development might have ceased and experienced serious setback. The meaning of the phrase quality of life has differed on good deal as it is variously used but, in general, it is intended to refer to either the conditions of the environment in which people live, (air and water pollution, or poor housing, for example), or to some attribute of people themselves (such as health or educational achievement). Central to this developing interest in quality of life is research follows into the relationship between people and their everyday environments.

Seeking to understand the nature of the *person–environment relationship* is the quintessential geographical question that lies at the core of the sub-discipline of social geography. In the specific context of the built environment, this can be interpreted as a concern with the degree of congruence or dissonance between city dwellers and their urban surroundings (Pacione, 2003; Michelson, 1997). Early studies of life quality were concerned mainly with defining the meaning of the concept (for example, Wingo, 1973; Hornback & Shaw, 1973). More recent investigations have advanced from pedagogic definition of quality of life to identification of the major components of the concept. In addressing this question there are two primary ways of determining the elements of life quality. The first is derived from the theories of, in psychology and sociology. The second method is to attempt to discover elements of quality of life by asking people directly how they view their own state of well being (Andrews & Withey, 1974).

The quality of life construct has gained momentum in human services over the last 50 years and therefore is a recent phenomenon that begins in 1960 and research on it developed in exponential rate (Cummins, 1997). It is a multilevel and amorphous concept that contains macro (income, employment, housing, education, living standard as well as environmental circumstances) and micro (perception of overall quality of life, individual experiences and value and other proxy indicators such as well-being, happiness and life satisfaction) approaches. Since mid 1980, there has been phenomenal growth in the research and measurement of concept of quality of life at macro, meso and micro level. The quality of life research is in its infancy in India. This concept is amorphous and vaguely defined (Campbell et al, 1976; Bowling, 1995; (Nussbaum & Sen, 1993; Rapley, 2003).

According to Handerson (2002) employment, income, shelter, infrastructure, energy, education, public safety, human right, environment, recreation, crime rate, and security from the disease are the indicators of quality of life. It may include nutrition, air quality, incidence of disease, crime rates, health care, educational services and divorce rate (McCall, 2008). Quality of life is, of course, predominantly determined by the individuals' health conditions, including the sensory system and cognitive functions, their functional level in daily life, coping recourses and available social support, their financial situation, environmental and community (Richter, Schwarz, & Bauer, 2008). Famous economists and philosopher Amartya Sen (1993) used capability approach beside other health related quality of life. The term *quality of life* is used to evaluate the general well-being of individuals and societies. The term is used in a wide range of contexts, including the fields of international development. In the words of Cummins (1997) the "*quality of life comprises both objective and subjective axis, therefore, should be the aggregate of seven domain such as material wellbeing, health, productivity, intimacy, safety, community and emotional wellbeing*".

According to World Health Organization the "quality of life is as an individuals perception of their position in life in the context of the culture and value system in which they live and in relation to their goal, expectations, standard and concerns. It is a broad ranging concept affected in complex way by the persons physical health, psychological status, level of independence, social relationship and their relationship to salient features of their environment "(Bowling, 1995, p. 3). For better quality of life it is becomes very important that person at physical, social, psychological and spiritual level be perfect. His mental or cognitive processes be in accordance with the environment as well as socio-cultural practices. Beside this the person be in a harmonious with his inside experience. The ego state model as developed by Eric Berne, the founder of transactional analysis also use to make sense of people's psychological or insides experience while adjusting to the society or forming interpersonal relationship. The ego state model says that we can understand our inside world as consisting of three different areas which are called the parent ego state, the adult ego state and the child ego state. We represent them as three circles, stacked upon each other with the parent ego state on top, the adult ego state in the middle, and the child ego state at the bottom.

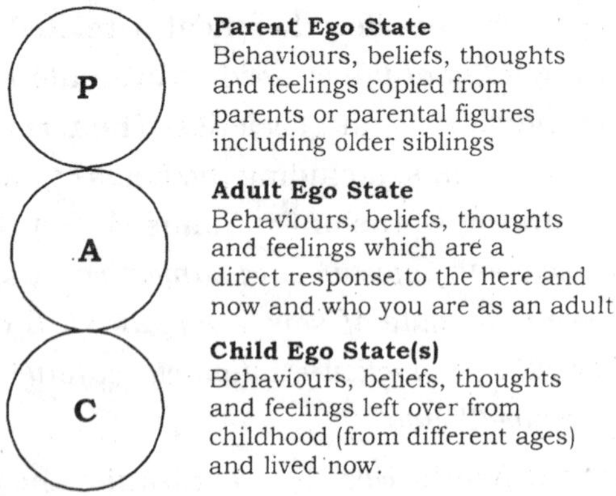

Adapted from Eric Berne (1961) Transactional Analysis in Psychotheraoy. New York: Grove Press

Fig. 2.2.1(a): Ego states model (Sourced: Eric Berne, 1961)

Each ego state is consistent within itself, so thinking feeling and behavior make sense and fit together within each ego state. However, between ego states there may not be any obvious consistency and they may contradict to each other. As a convention we abbreviate parent, adult and child ego state to Parent, Adult and Child and write them with capitals to make it clear that we mean ego states and not people. The ego state model gives our internal psychology a historic perspective: under stress we move into old patterns, either our parent's patterns (parent ego state) or the ones we had as children (child ego state). The parent ego state may include different sets of representations - one for mother, one for dad and possibly others for other people. Our child ego state will represent our historic development, so we may have a strong layer at age 2, lots of stuff from age 6, and some leftover issues from our teenage years, to mention but a few. More appropriately one ego states are directly related to quality of life.

The quality of Life is an area of study that has attracted an ever-increasing amount of interest over the past two decades, particularly in the areas of health, rehabilitation, disabilities studies, and social services, but also in medicine, education, and others. The study of Quality of Life is an examination of influences upon the goodness and meaning in life, as well as people's happiness and well-being. The potential of the quality of life (QoL) concept lies in its positive meaning and interdisciplinary acceptance (Lindström, 2005). It involves degree of capacity or first and good capacity life as well as having, loving and being concept as well as material and immaterial resources. More appropriately this concept is quite complex that may include objective indicator (e.g. wealth); subjective indicator (value and subjective perception)l; satisfaction of human need;

psychological growth (e.g. cognitive competence); health and functioning indicator; social health indicator (societal adjustment); social cohesion and social capital indicator; environmental, hermeneutic (Individual values) indicator etc.

2.2 POLYGAMOUS LIFE STYLE IN TRIBAL AREAS

Practitioners and proponents of polyandry offer several reasons for the continuation of this practice, including polyandry's enhancement of biological fitness in offspring, the abundance of available marriageable men compared to women, polyandry's economic benefits within the society, and the perpetuation of a polyandrous tradition established in the Hindu epic the Mahabharata.

Fig: 2.2.2 (b): Marriage customs in Kinnaur (Janekang)

Polyandry, the marriage of a woman to multiple husbands, is first recounted not in historical texts, but in the Hindu epic the Mahabharata. In the Mahabharata, the princess Draupadi is married to the five heroic Pandava brothers, and the Pahari of Himalayan India have continued this practice of fraternal polyandry through the centuries. Paharis, or mountain people, are a separate culture of India characterized by several diversions from mainstream Hinduism, including the institution of polyandry, and a separate linguistic dialect. Their centuries-old trading partners to the north, the Tibetan Nepali, also practice this rare marriage form, which is generally believed to have been introduced to them by the Pahari (Berreman, 1962).

Fig. 2.2.3 (c): Traditional dress in marriage at marriage in (H.P)

Polyandry is typically used to refer to the practice of male-initiated marriage and sexual relations with already-married women. As a definitional matter, "polyandry" need not be limited to male-initiated bonds, and in some polyandrous societies, women have freedom to choose their mates. However, Mormon polyandry as practiced in Nauvoo (and later to a lesser degree in Utah) did not accord women such freedom. Rather, a small subset of already-married women were approached and asked to become the plural wives of other men, while simultaneously remaining in their existing marriages. In a polyandrous system - particularly a system of male-initiated polyandry as was practiced in the early church Under polyandry, the costs of spouse-sharing are distributed at least somewhat among both sexes. Second, polyandry probably fosters male empathy towards women forced to share spouses.Polyandry in Tibet (also practiced in other Himalayan regions) is a form of polygamy whereby a woman has several husbands. In Tibet traditionally those husbands are brothers, which is why it is most commonly called "Fraternal Polyandry" and concern over which children are fathered by which brother is discouraged. Historically, the social system also compelled everyone to abide by the tradition of marrying within their own social class, and also among their own blood relations. Because Tibet who owned property - by not allowing it to be split between brothers.

The polygamous marriage therefore arose as a solution to this potential threat. Records of polyandry as practiced suggest that it did not afford greater relationship choices to women, but rather created extra layers of angst and heartache within their existing relationships. In theory, polyandry could be a version of polygamy that is more feminist-friendly than classic polygyny. As noted above, it removes some of the double standard, and it seems to create — again, in theory — additional opportunities for women to enter into rewarding relationships. However, the history of polyandry as practiced, and the relationship limitations imposed under a system of male-initiated polyandry, weigh more heavily on the other side of the balance. Polyandry would not be a more feminist version of polygamy; in practice, it would almost certainly be more anti-feminist. There aren't very many institutions that can make a culture of male-initiated polygyny look good by comparison — but a male-initiated system of polyandry is probably one of them.

Temporal spiritual polyandry is still practiced today in tribal culture. Widows *or* women who have legal divorces (but not temple cancellations) who remarry in a civil ceremony or in a temporal temple ceremony are technically married to two men at once. If children are born through a second husband, they are technically sealed to the first husband. This creates an 'ownership' question. Consequently, widows or divorcees (without temple cancellations) have *more* of a disadvantage in seeking future husbands than men who are dating and have previous wives. Men who marry a second wife offer the second wife complete co-ownership of the children, whereas women who marry a second husband cannot. While our pioneer ancestors were seemingly less shaken by this, it does seem to be an issue on the LDS single adult dating circuit today. Polyandry introduces a lot of potential benefits for women — greater power within their relationships; greater ability to enter into rewarding and fulfilling relationships; and (if put into place with polygyny) a measure of gender equality in the polygamy sphere.

The tribal women, as Elwin (1958) believes is exactly the same as any other women, with the same passions, love and fears, the same devotion to the home, to husband and children, the same faults, and the same duties. In practice, however, the tribal women enjoy comparatively more freedom and are devoid of many constraints imposed women in non-tribal societies (Hodson, 1975; Mann, 1996). Tribal in transition however, stand somewhere in between, but even here the traditional norms are somewhat liberal to womanhood. The least influenced by formal legislation is the tribal women. She continues to live in her customary style, with the same position and behavior on the one hand, though she leads a hard life. The tribal women enjoys far more freedom than her caste counterpart. She can roam about freely, visit markets or friends, cut jokes with them without reservation and is usually free to select her mate. Some part of her freedom gets restrained in past restrained in post marital style, although she has the liberty to divorce and remarry within the tribal set up, a divorced women is not much stigmatized. The bride price is said to be a mark of respect and value for her. She has a firm hand in most of the family matters. Her social participation is not as aggressively restrained as in the case of caste women. On the other hand, however, in most patriarchal tribal communities, she cannot customarily speaking became a religions or a political head. She is deprived even of family headships and property inheritance.

Fig: 2.2.4 (d): Kinnauri women and men attending marriage party

The Pahari people offer another reason for the continuation of fraternal polyandry: a particular spiritual devotion to Draupadi and her husbands in the Mahabharata. The Mahabharata maintains that the Pandavas traversed Pahari lands during their wanderings and Paharis claim that polyandry is conducted in traditional remembrance of these heroes. However, the Mahabharata is a poor source in which to find the virtues of polyandry. Draupadi is the only woman of the epic involved with a polyandrous marriage; her husbands are the sons of a polygynous union, and, in fact, each of her husbands maintains at least one personal wife of his own. Paharis find sexual access between the wives and brothers of monogamous marriages as acceptable and desirable as the legitimized union of brothers in polyandry, but the individual wives of the Pandavas would be looked upon with disfavor, as they are not shared equally among the brothers (Berreman, 1962). Thus, the emulation of the Pandavas through polyandry is an insufficient explanation of its presence, since not all marriage traits of the Pandavas are equally respected by the Pahari. Thus, no solution provided by either the peoples who practice polyandry or varied anthropological hypotheses on the subject succinctly and completely explains its continued practice.

2.3 EVOLUTION OF REET SYSTEM IN INDIAN INSTITUTIONS

The evolution of social institutions is generally considered from the time of Mauryas to the fall of Kushanas. There exists heterogeneity admixture of ethnological or racilal elements in India. The population here is generally divided into four ethnological groups. The first group is named as Negritos or Negroid race those came from Africa along coast lands of Arabia and Iran, settled in Western and Southern India passed over to Malaya and islands of Indonesia. Most of them were killed and some rare escaped and now a days found in Bihar. Second is Proto-Australoids or Austric people come from West or Palestine and spred in East India. They are aborigines of Australia. They are known as forefather of lower castes such as Santhal, Kols, Bhils, Mundas etc those have wavy, curly hair, broad nose and darkish skin. Brahamputra River found in Assam, Bhutan and Nepal. They were known as Kiratas according to Manu falls in Vratyas of Kshatriya castes. In Mahabharata they are known as Mlecchas as sinful or barbarous people of North. They are also known with other name such as Yavanas, Kambojas, Barbars etc.

Fourth group is Dravadian came from East of Mediterranean region from Asia and settled in Mesopotamia and Persia, Iraq and Iran. They were forerunner of Aryan culture. There were two categories of Aryan people Aryan with white texture with better economy and another were Dasas or Dasyus. They fight themselves to show their supreme power and Aryan defeated Dasas or Dasyus and defeated the black skins and ugly look Dasyus . On the basis of their colour and racial. The lower castes people are in our country belongs to Dasas or Dasyus,Dahai or Dahya. According to Purusa Sukta of Rigveda, especially in Dharma sutra and Dharma sastra lilterature there were four Varnas in Aryan society named as Brahmanas, Kshatriyas, Vaisyas and Sudras. The fourth group was considered untouchable by Patanjali (Mukerjee, 1976, p. 17) in India whose origin are from Dasyus or is named as Sudra as originated from Sodre or Sogdi inhabiting in Northern Arachosia. Since then the untouchability system is found in India in general as well as in other parts of the country.

There are various polyandrous caste hierarchies in India, in tribal and in Himachal Pradesh those includes Bhat, Brahaman (adopt janeo or sacred thread), Deva, Dethi, Kanet (Yeomen of the state), Rajpur (Khash), Koli, Lohar, Badi, Bajgi, Chanal, Dhaki/Turi and Dumra. Bhat Kanet and Deva can enter temple and are up in the social ladder. There are also Kanets, Rathis, Rawats, Negis and Thakurs hold in similar status. The Brahaman are general purohit and cooks abstain from manual labour. Some other rahaman also work as *Hal-bahu*. Pabucha brahaman of Sirmour also are involved in inter-caste marriages between upper castes (Bhat and Kanets) whereas they do not marry with lower castes. They also adopt Karewa (widow marriage) system. Rajpur or Kanet wish them Palagan i.e. paon lagan (I touch your feet) and brahaman give ashirvad or blessing, demon, and witchcraft. Bhat is one who can cure people of the influence or pariyan (evil spirits) and practices the Tantric vidya or black magic, jantras and mantras. Earlier Kolis and Dhakis were expert in the field later Rajputs, Kanets and Brahaman adopted this system from them. There are also Koli bhat. Kanet who serve as a priest is called Negi who should always marry with Negi girls. But the Negis in Himachal Pradesh are called for Kinnauri people. The Dethi stand lowest among the Brahaman who are originally Charjas (perform religious ceremonies of the dead only).

In fact they belongs to Aryan society who probably passed through a stage in which promiscuity was tolerated as a necessary evil, if not approved. The same account states that the term Kunit was first applied to Brahamans and Rajputs who in a strange country peopled by a primitive race, abandoned the orthodox tenets of Hinduism and lapsed into such practices as widow remarriage. The Kunits were violators and Rajputs or Chatris were Kunit or Kanet derived from Kania-het (love of daughters), raja Kans of Puranas or Krishna itself. According to Alexander Cunningham, the Kanets were the descendants of Kunindas or Kulindas, rich and powerful people who ruled about 15BC between Beas and Tons rivers and were the original inhabitants of the lower slopes.

The Reet is practiced by upper and lower level caste in Himachal region. Polyandry can not be understood without an adequate comprehension of custom of Reet. The Reet permits a woman to secure release from one set of husbands to may another. This custom ensures that each set of brothers would get, for some time at least, a common wife although there is no guarantee that she would stay with them permanently. Thus Reet extends the extent of sharing of scarce women by larger groups. Polyandry enjoins that brothers have a common wife or common wives if they can afford them. Reet permits the movement of women from one set of brothers to another in the course of her several marital unions after customary divorce. In a society which regards women as a chattel – a commodity to be bought and sold, the practice of Reet permits her release from inhospitable domestic environments and serves to demonstrate her value. There is, of course, the possibility of the custom being abused and of families disposing of women for temporary economic gain even tough the women themselves may not desire such a move. The women are so highly prized that Reet unions do not in any manner cause any stigma to them. They in fact put their price up (Dube, 1975).

According to Sirmur State Gazetteer, if a person marries one of a lower caste he is expelled from his caste, but if he merely cohabits with one of the lower caste he is excused.

The miscreant is outclassed and fined. The person who eats, drink and smoke with him are also considered as outcastes. For punishment as well as purification Panchayatas are called who purify them by advising to bath in Ganges and perform a prayaschit (peance) or to present he goat to them. There is general rule that the upper caste person should not marry with lower caste and nor he should enter and live in the house of lower caste and nor he should eat. The sexual relation with lowest caste (e.g. Chamar) is not tolerated. Person if found so they are excommunicated. But with some castes e.g. intimacy with Turi women is acceptable provided that the upper caste person does not eat and drink with her. High castes in Himachal Pradesh generally forms good families.

Fig: 2.2.5 (e): System of Marital Reet in Kinnaur District (H. P.)

Providing brides to Bhats and Kanets, the bhat girls were also given to the members of the then ruling family and their children called Kanwars. The girls were given with the bride as Khawas or dasi (slave girls). Kanet girls also were taken as Khawas in this manner as well as by payment of Reet. Their children, designated as Sarteras could not inherit property, only were given maintenance at the sweet will of the father although they were Kanwars yet as unfortunate class who did not have slight right for the properties. In upper Bahaahr Brahamans and Rajputs marry Kanet girls but their children are considered as Sarteras who gain status of his father after three or four generation. A raja or a rana was succeeded by a Sartera son and latters descendants had in course of time been accepted in marriage by Rajput families. In Dravidian kinship terminologies, there is widespread practice of cross-cousin marriage. Beside this there is a single term for wife's father, mother's brother and father's sister's husband in the Toda kinship vocabulary because by the operation of the rule of bilateral cross-cousin marriage, this might well be one and the same person.

The structure of kinship terminologies reflects crucial features of the family and marital arrangements that exist or at least had once existed among these peoples and he further speculated that on this evidence societies could be arranged in a multi-stage evolutionary series, culminating in the monogamous family of present day Europe. Joint family is the source of Indian social backwardness and economic stagnation, creating psychological dependence, stifling the initiative of individuals, militating against capital accumulation and rational economic development (for an account and critique of these formulations). It is of course commonplace to remark that India is a Plural or multi-cultural society and a highly stratified one besides, but the domain of kinship is an area in which this is patently evident. Hindu joint family was first given systematic shape in the writings of Sir Henry Maine, who believed that he had discovered in India a living example of the patriarchal family of ancient times. Maine had also proposed a theory of social change which linked stages of human progress with changing forms and functions of the family.

Though the question of the Unity and the variety of Indian kinship systems is not one of special concern in the accepted academic paradigms of the discipline. In fact it is one that has preoccupied scholars of Indian family, kinship and marriages. The first one deals with the qualitative dimension of love, sex, marriage and family life as sociologists, we have on the whole fought for chronicling the emotional tenor of family relations. Marriage is almost universally an occasion for material transactions, It is also important to see how these exchanges map on to the logic of the marital exchange. Some fragmentary possibilities for a woman-centered view on Indian family and kinship, but they do not of course amount to a gender-sensitive rewriting of this discursive field. Indian kinship systems seem to be extraordinarily heterogeneous. They are regionally diverse (north/south/east/west), communally differentiated (Hindu/Muslim/Sikh/ Christian/Buddhist etc.), socially stratified (in terms of case of class) and culturally discrete (by ethno-linguistic criteria). In fact, there seem to be as many varieties of kinship organization as there are distinct communities.

Irawati Karve identified three major types of kinship organization-in her terms, Northern, Southern and Eastern co-ordinate with the culture areas of the three major language families (the Indo Aryan or Sanskritic, the Dravidian and the Austro-Asiatic). Alongwith between the northern and southern types and combining features of both. According to Thomas Trautmann, on the other hand, the historical convergence that has taken place (particularly in the central zone) between the Dravidian and the Indo-Aryan kinship system is the outcome of the political and ideological domination of the north over the south through the millennia. Specifically he proposes that the ideology of *kanyadana,* the unrequited 'gift of a virgin" which is the central motif of Hindu marriage, has worked to skew an original Dravidian marriage rule of symmetrical sister exchange in the direction of unilaterality, since it accords *'wife-takers'* superior status over *'wife-givers.* It is the principle of *kanyadana,* he maintains, that truly expresses the uniquely Indian culture of kinship.

North Indian marriage, which at first glance appears to be determined only by the rules of ***sapinda*** exogamy and caste endogamy, can be best understood as an alliance system in which 'caste' plays the part of the unilateral preferences of the Dravidian marriage system. In the hypergamous milieu of north India, wife-givers and wife-takers may never be confused, but hierarchical relations of alliance are reiterated over time, and the affinal obligations in respect of ceremonial functions and gift-giving are 'inherited' in a manner similar to that described for sought India "Affinity" in a value in north India, too. Matrilateral cross-cousin marriage entails the reiteration of wife-giving and wife-receiving relations in successive generations, every daughter following the alliance pattern of her father's sister. Hypergamy (where women marry into higher status groups) and hypogamy, where women marry men of lower social status, also falls were the Reet system is identified custom in the Himalayan conditions and Himachal Pradesh. Similarly the polyandry has various theoretical perspectives. The discrimination as follows:

2.4 THEORIES OF POLYANDRY

2.4.1 General Theory

It states that the polyandry is the insurance of marriage for all eligible individuals in societies where there is a surplus of men (Heath, 1955; Berreman 1962). In Jaunsar Bawar, a polyandrous Pahari community, there is an expectantly greater shortage of women. However, in nearby and monogamous Garhwal, there is a high surplus of women (Berreman 1962), a trend which suggests that polyandry may be the cause and not the solution to a shortage of available wives. Additionally, in Tibetan Nepali communities, there remain single women who are unable to find husbands because of their status as daughters of an illegitimate union (**Schuler, 1987**). Indeed, these single women are ill-favored candidates for marriage in a culture which values the wealth associated with legitimacy. Single women are part of a peripheral population which is a concomitant of the system of social stratification, of the narrow principles governing entitlement to the status and property (**Schuler**, 1987). Although illegitimacy is not necessarily stigmatized by Tibetan societies, yet it does preclude the usual inheritance of wealth, making illegitimate daughters less likely to be married into a polyandrous union. This particular social eligibility of women and not their sheer numerical representation is a more accurate factor of the ratio between available men and women in a polyandrous society.

The economic benefits in a system of familial partition are another reason. By far the most convincing explanation of polyandry lays in the standard division of familial property among both Pahari and Tibetan Nepali societies. Both societies practice familial partition, in which land holdings and resources are divided more or less equally among all legitimate sons, with smaller land holdings and resources also available to illegitimate sons. Among the Nepali, this particular inheritance is classified as part of the *trongba* household and a percentage of the family's inheritance is especially bestowed on the eldest son. In the case of partition, sons form their own *trongba* households and, if they gain prestige and wealth with their resources, are granted individual lineage names (Levine 1988).With a decreasing availability of resources, the ideal way for a family to retain the wealth of their lands is to keep the lands intact. In order to mitigate the partition of land, brothers are reunited under one household by marriage to a common wife. The sons of their polyandrous union are also expected to marry in polyandry, thus retaining the entirety of family land holdings through generations.

Indeed, the wealthiest of three Tibetan Nepali communities, *Ladog,* is also the one which practices polyandry most consistently, where not only do more women marry brothers, but also marry more brothers within the same family (Levine 1987). Ideal as polyandry may seem to the system of partition, it is not the only form of marriage practiced in either the Pahari region or in Tibetan Nepal. Even in *Ladog*, 34.6% of the women marry monogamously and another 6.9% of marriages are monogamous marriages resulting from a brother leaving a polyandrous union (Levine, 1987). In fact, at the time of partitioning between brothers who leave the union, harmonious familial relationships are often jeopardized as "differential paternity that becomes a public issue" (Levine, 1987:274). The brothers, formerly the conjoint fathers of all their wife's children openly acknowledge their particular paternities. In the case of illegitimate children, the responsibility of which is "socially enforced," partitioning becomes an even more contentious debate as the rights of illegitimate children threaten the property rights of legitimate ones (Levine, 1987:279). Thus, the economic benefits of polyandry are often outweighed by the number of brothers who partition land from the family in monogamous unions and the divisive effect of such partitioning among brothers who chooses polyandry as Hindu tradition

2.4.2 Enhancement of Biological Fitness Theory

Darwinian fitness is the product of two variables. First viability that is defined as the probability of survival to reproduce. Second deals with and fecundity that is defined as the number of offspring produced. Polyandry does not empirically increase the biological fitness of offspring within a marriage. Although polyandrous marriages do ensure some variation of alleles within a familial population, the probability of transference of alleles from father to children decreases with the number of husbands within a polyandrous marriage. In a polyandrous marriage of only two brothers, the opportunity of a husband to pass half his genetic code to his children is already decreased by half. This, of course, assumes that all husbands are provided with equal sexual access to the common wife, a cultural ideal which is not universally maintained (Levine, 1987). In the case of unequal access, the probability of allele transferal remains reduced for all husbands and is even more significantly decreased for the husbands provided with less sexual access. Additionally, the marriage of multiple husbands does not increase the number of offspring born to a common wife. Plausibly, the common wife might be expected to raise more children to adulthood than a monogamous wife due to her increased sexual access and ensured providing by one or more husbands. Polyandrous marriages, however, produce the same amount of offspring as monogamous marriages

– due to the universal gestation periods of women – and in fact have a higher offspring mortality rate resulting in less surviving children. The decrease in the number of offspring further decreases the chance of transferal of alleles to a further generation. Thus, polyandry "does not appear to enhance the fitness of individuals who practice it, and in fact, seems to entail substantial reproductive sacrifice". Polyandry does not increase the biological fitness of offspring; in fact, it decreases the likelihood of transferring desired traits and alleles. The surplus of men in a polyandrous society may be the result of polyandrous marriages instead of the solution, and many single women remain who are considered ineligible for marriage by other social standards. In the system of partitioning, in which complete and continued practice of polyandry is the most sensible means of retaining family land holdings, the total participation of brothers in polyandrous marriages is rare and land holdings are still often divided amongst families. Finally, the same epic text that provides the traditional foundation for polyandry also provides the foundation for an inequality of marriages among brothers and polygyny, neither of which is practiced to the same extent as polyandry in either Tibetan Nepal or the Pahari region. Lack of substantial support for polyandry demonstrates its anomaly among marriage institutions of the world and further studies among other polyandrous cultures in India and Nepal are necessary to discover the benefits of polyandry in a contemporary society.

2.4.3 Conflict Theory of Evolution of Imprinting

Genomic imprinting is the parental germline-specific epigenetic modification and subsequent differential expression of genes during embryonic and postnatal mammalian development and during plant endosperm development (John & Surani, ; Baroux *et al.*). In mammals, a large proportion of imprinted genes affect fetal and placental growth, with expressed paternal alleles generally increasing growth and expressed maternal alleles decreasing growth (Beechey, 1999; Moore, ; Reik *et al.* 2001; Wilkins & Haig). The "conflict" theory of evolution of imprinting provides an explanation for the parental origin-dependent gene expression levels in terms of relatedness asymmetries between the parental alleles at loci that, when expressed in embryonic tissues, influence the amount of resources received from the mother by her offspring (Haig & Westoby, Moore & Haig,). Different aspects of the evolution of imprinting have been modeled by a variety of methods. Much of this work has been extensively reviewed elsewhere (Haig, Spencer ; Wilkins & Haig). In relation to models that provide explicit tests of the conflict theory of imprinting, two approaches have predominated. Haig and colleagues (Haig, Wilkins & Haig) used game theoretical models, which seek to define an unbeatable, evolutionary stable state (ESS) at equilibrium. These models found that imprinting evolves exclusively under polyandry, with paternally inherited alleles favoring increased maternal investment in offspring, and maternally inherited alleles favoring reduced investment, consistent with the predictions of the theory. Mochizuki *et al.* (1996) reached at similar conclusions using multivariate quantitative genetic models, which defined, at equilibrium, stable values of mean population gene expression levels at zygotic loci that influence maternal investment in offspring. In addition, Mochizuki *et al.* , (1996) attempted to explain other aspects of imprinted gene biology in their models; for example, the perceived paucity of imprinted genes was explained in terms of selection against recessive deleterious mutations in their coding regions. However, deleterious mutations are unlikely to provide a sufficiently strong selective force to prevent the evolution of imprinting (Haig, & Spencer), and other more plausible explanations have been proposed (Haig)

2.4.5 Psychological Theory

The psychological theory predicated on the ridiculous presumption that only men really enjoy and crave sex. The Sex addiction is simply a new name for the old evolutionary concept—the innate urge to impregnate as many females as possible. Unlike addictions to alcohol, cocaine, and cigarettes, in which the craving is induced by external elements, sexual craving, by its nature, is an innate and natural phenomenon. The sex addiction is a specific situation—the frequency of erection and the intensity of orgasm—dependent on the person's blood-level and testosterone. Man deals according to his physical, social, mental and financial conditions. Perhaps the women might not want to castrate their husbands, therefore solely depends on male.

2.5 MERITS AND DEMERITS OF POLYGAMY

Society assumes a male sexuality only on one mode, and that's masturbatory, and that sex with women is only better because of the physical sensations, and not because of women's personalities and presence can be enjoyable in and of itself. And that tenderness shown to women is an act where men put on out of chivalry, not out of genuine feelings. Karl Marx and pioneer in communist philosophy, wrote about monogamous marriage in his book, *The Origin of the Family, Private Property, and the State*. Engels describes monogamous marriage as a social institution designed for two main functions. First, monogamous marriage ensured wealth was passed down to biologically related offspring. Second, monogamous marriage trapped women in a life of unpaid domestic and childrearing labor. Engels believed that the communist revolution would undermine both of these functions. A communist society would no longer allow wealth to be passed down to biological offspring. Therefore a communist society would socialize the work of raising children. The monogamous marriage would no longer serve any purpose in communist society. According to Engels, the rise of monogamous marriage coincide with oppression of women by the men.

Generally it is believed that the monogamous marriage is much better as compared to the plural or polygamous marriage. Certain scholar pinpoints the default in monogamy and recommended it to marry more than one wife. The men are generally considered as with stronger sexual drive as compared to women. Some cultures value monogamy as an ideal form of family organization. The "mono-marital principle" dictates that for each and every generation, one and only one marriage is permitted collectively among all the male siblings, and the children born out of this marriage are members of the family unit who have full legal rights. The family organization was based on these two patterns to avoid the partitioning of their estates. A generation with two or more conjugal families was seen as unstable because it could produce serious conflicts that may divide their corporate family land. As a matter of the fact, Tibetan inheritance rules of family land, mainly based on agnatic links, did provide for each generation to partition the land between brothers, but this was ignored to prevent the estate unit from being threatened.

However, many cultures prefer other forms of family organization. Anthropological data suggests that a majority of societies prefer polygamous marriage as a cultural ideal. There are multiple forms of non-monogamy that are used to organize families, as well multiple forms of monogamy such as marriage, cohabitation and extended families. People disagree strongly about the value of monogamy that has been criticized as well as supported. Two common criticisms of monogamy are that socially monogamous marriage oppresses women and that lifelong sexual monogamy is unrealistic. Supporters of monogamy have argued that a society that supports monogamous marriage can promote women's equality and that sexual monogamy facilitates intimate and lasting relationships.

There is coercion with the female in polygamous marriage. In this coercion, the women are compelled by a male to involuntarily behave in a certain way (whether through action or inaction) by use of threats, intimidation or some other form of pressure or force. It typically involves the actual infliction of physical or psychological harm in order to enhance the credibility of a threat. Physical coercion is the most commonly considered form, where the content of the conditional threat is the use of force against the person, the dear ones or the property of the victim. An oft-used example is "putting a gun to someone's head" to compel action. In psychological coercion, the threatened injury regards the victim's relationships with other people. The most obvious example is emotional *blackmailing* through thought or language. In cultural coercion the women are forced to wear certain style of dress, publicly reciting a creed or a pledge of allegiance they find ethically reprehensible, starting to smoke when they'd rather not, etc. Economic coercion is practiced when a controller of a vital resource uses his advantage to compel a person to do something he would not do if this resource were not monopolized. Religious coercion is the most ancient, extensive and durable kind of thought coercion has concerned religion in which the selfish entity is a supernatural one.

Absent the threat of polyandry, it is easy for men to sit around and blithely discuss the need for wives to share a husband — after all; men reap all of the benefits and feel none of the loss. It is quite another thing for men to be forced to internalize these questions. In a polyandrous society, the analysis for men moves away from "women should be happy to share their spouses" and becomes a self-query of "would I be happy to share my spouse?" Polyandry thus forces men to internalize some of the costs of polygamy, rather then offloading all of the costs onto women. Third, polyandry potentially allows more freedom for women to enter into fulfilling relationships. Mormon polyandry, as practiced, may not appear to grant this benefit. (This is because all polyandrous marriages were male-initiated and mostly initiated by a small group of high-level church leaders). However, it seems likely that even in a society where only male-initiated polyandry were allowed, women could use informal means to enter into relationships that they felt were desirable. That is, a woman could approach a desired suitor and suggest that he request her hand as a polyandrous bride. Thus, polyandry could become a potential avenue for women to seek and maintain fulfilling relationships with desired partners.

A compelling socio-biological justification for polyandry is that it makes good genetic sense for brothers to raise one another's children since a brother possesses the next closest gene pool to their own. The Tibetan social organization under Lhasa control from the 17th century on was quasi-feudal, in that arable land was divided and owned by aristocratic families, religious organizations, and the central government and the population was subject to those district divisions. The population was further divided into social classes: aristocratic lords (*ger-ba*); monastics (as much as 20% of the population; subjects (*mi-ser*) ; consisting of tax payer families (*tre-ba* or *khral-pa*); householders (*du-jong* or *dud-chung-ba*) and landless peasants.

FAMILY ENVIRONMENT AND EMOTIONAL WELLBEING

"Love can only be exchanged for love and trust for trust. If you love without evoking love in return you are not able by the manifestation of yourself as a loving person to make yourself a beloved person then your love is impotent and a misfortune. If you wish to influence other people you must be a person who really has a stimulating and encouraging effect upon other"

–Carl Marx

3.1 MARRIAGE, MENTAL STATES AND HAPPINESS

It becomes a matter of an exploration whether there are any link between the marriage and mental health. Marital status as a network indicator has shown married people to have better physical and mental health and also higher survival than those who have never married, widows, widowers and the divorcees (Gove, 1973, MacIntyre, 1986; Helsing & Szklo, 1981). The benefits of marriage always appear to be greater for men than the women (House & Kahn, 1985). Evidence linking psychiatric impairment with divorced marital status and the presence of marital discord is reviewed.

Various theoretical models to explain these relationships are considered. Divorced marital status and marital discord are related to psychiatric impairment and mental health service utilization (Segraves, 1980). Different explanations have been given such as the stress of bereavement, lack of social support, role change and artifact effects.

Fig. 3.1.1 (a): A perfect photo of traditional Kinnauri marriage

In an extensive review of the literature, House, Landis and Umberson summarized the present situation as follows: "Prospective studies, which control for baseline health status, consistently show increased risk of death among persons with low quantity and sometime low quality, of social relationships. Experimental and quasi-experimental studies of humans and animals suggest that social isolation is a major risk factor for mortality from widely varying causes. House et al (1988) also indicated that the risks associated with social relationships, rival risk factors such as cigarette smoking, high blood pressure, elevated blood lipids, obesity and physical inactivity, are threat to wellbeing. The mechanisms however, through which the network variables influence health remain unclear (Modi, 2001), need further exploration.

In a study Myers (1992) reviewed literature to establish the link between growth and mental development. He concluded that satisfying psychosocial needs can have an effect on nutritional status through its effect on metabolism linked to stress reduction, and by helping to produce changes in the care demanded and provided. At the same time, nutrition is seen to have an effect on psychosocial development, operating primarily through its impact on attention, responsiveness, independence, irritability, and affect. Nutrition is one of the complex factors operating to influence development and associated behavior. Women with polygamous marriage report more psychological distress, high level of somatization, phobia and other psychological problem as well as less life satisfaction.

Emotional wellbeing and mental health concerns are major health issues in their own right. They can also greatly affect physical health. The married people are less depressed with fewer alcohol related problems, as compared to unmarried people. The rate of decrease in alcohol problems and depression is faster in married people, than in unmarried people. Married men report lesser cases of depression, and married women report fewer cases of alcohol related problems. Better mental health fails to show any relationship with length of marriage and presence of children. The quality of marriage may be responsible for the relationship between mental health and marriage . Further a research on the association of marital status and happiness was conducted in 17 industrialized nations. Results show that happiness is significantly higher in married individuals as compared to the unmarried persons because of the increase in financial satisfaction and high levels of perceived health during marriage. Married couples are also happier than couples who are just cohabitating as the latter are lacking in financial and health protection that can only be given by marriage (Steven, & Ross, 1998).

Study of Al-krenawi, Graham, and Slonim-nevo (2002) consider the mental health consequences of polygamy in a sample of 101 Arab Muslim adolescents (19 from polygamous and 82 from monogamous families) at Juarish (Ramla), Israel. The respondents completed the Self-Esteem Scale (Rosenberg, 1979), the Brief Symptom Inventory (Derogatis & Melsavados, 1983; Derogatis & Spencer, 1982), and the McMaster Family Assessment Device (Epstein, Baldwin, & Bishop, 1983). The respondents from polygamous families had lower self- esteem scores, statistically significant higher scores in 2 BSI dimensions, higher scores in all other brief symptom dimensions, and higher levels of self-reported family dysfunction. The respondents from polygamous families reported lower levels of socioeconomic status, academic achievement, and parental academic attainment. These variables may have had a more direct impact on mental health than did parental marital status. The data also indicates that perceived family functioning was the best predictor of mental health

Getting back into a relationship after a marriage fails is good for men's mental health, but bad for women's, suggest the results of national survey in the U.K. The study also found that cohabiting is better for men's mental health, but marriage is better for women. The research is based on responses to a validated mental health questionnaire from 4,430 men and women under the age of 65, part of the annual British Household Panel Survey, which began in

1991. Men whose relationships with a first partner fell apart had much poorer mental health than men who remained with their first partner. And those who decided to live with a new partner after a marriage break-up also had better mental health than men who stayed single or who remarried.

The mental health of women who had not split up from their partners was also better than that of women who did. But women's mental health progressively deteriorated the more break-ups they experienced and the more times they moved on to other relationships. Women who stayed single actually enjoyed much better mental health, which was not true of men. While the mental health of both men and women were better in long-term relationships. Men who chose to marry their partners fared had poorer mental health than those who chose just to live with them. Women fared better if they married. Living with a partner is better for mental health, but that women have a harder time emotionally when relationships fail (Holmes, 2004).

The Harvard Health Publications' special health reports include reliable, authoritative mental health information on a wide range of issues that impact emotional well-being, along with treatment strategies and practical solutions to help you handle life's many challenges with resilience and maintain your emotional health. According to a major international study across 15 countries and 34,493 people, getting married is positive for the mental health of both men and women, resulting in reduced risks of the likelihood of most mental disorders such as depression, anxiety and substance abuse. By contrast, ending marriage through separation, divorce or being widowed, is associated with substantially increased risk of mental health disorders in both genders; particularly substance abuse for women and depression for men.

The wide-ranging study led by clinical psychologist Dr Kate Scott from the University of Otago Wellington is based on the WHO and World Mental Health (WMH) surveys across developing and developed countries in the past decade. There have been a number of international studies about the impact of marriage on the mental health of men and women and this is quite a controversial area because of the gender politics involved," says Dr Scott. "But what makes this investigation unique and more robust is the sample is so large and across so many countries and the fact that we have data not only on depression, which has been much studied in the past, but also on anxiety and substance use disorders. In addition we were able to look at what happens to mental health in marriage, both in comparison with never getting married, and with ending marriage. One of the more important findings is that in recent years it has been asserted that marriage is better for men than for women in terms of mental health. This study does not agree with that position. Result revealed that that never getting married is good for both men and women in terms of most mental health disorders. Other studies based on the WMH surveys have shown that as gender roles have become less traditional over time, for example as women have worked more and become better edučated, female depression tends to fall. The other study found that getting married reduces risk of substance use disorders more for women than for men. This may be explained by the fact that women are usually the primary caregiver for young children.

A number of international studies have shown that women's consumption of alcohol drops sharply when they become pregnant, and this restraint often continues into early child care. The marriage can increase the risk of mental health problems. Being previously married (i.e. being separated, divorced, and widowed) is associated with increased risk of all mental health disorders in both men and women; particularly with depression in men and substance abuse (drugs and alcohol) in women. The marriage can make people vulnerable to developing mental disorders (Ancell, 2009). Perhaps the mental health depends on the congenial atmosphere of the family.

3.2 FAMILY ENVIRONMENT AND EMOTIONAL WELLBEING

Living independently before marriage is part of a broader pattern of family and demographic change characterizing modern societies since WWII. A study was conducted by Shirley (1986) in order to assess the characteristics of 42 healthy single-parent families. A total of 84 subjects mean age 41.6 yrs and a target child (mean age 14.1 yrs) participated. The variables included were socioeconomic status (SES), social support, communication, religiousness, problem solving, and the physical and mental health status of single parents and their children. The effects of the sex of custodial parents and the custody arrangements on health outcomes were also analyzed. A multi-method, multivariable approach was used. The family environment scale, family interaction schedule and interview were used. Result revealed that the single parents and their children reported fairly high levels of both physical and mental health. Communication, social support, socio economic status, religiousness, and problem solving were also correlated with the mental and physical health of parents and children (Shirley, 1986).

A number of adolescent characteristics are examined to determine their impact on adolescent perceptions of the parental power structure in the family. The variables include adolescents' sex, grade, religiosity, and birth order. Relationships were explored by using "parental power" as a multidimensional variable, including perceived outcome-control, referent, legitimate, and expert power. Purposive sub samples were taken from a larger pool of high school and College adolescent yielding 458 questionnaires (231 Ms, 227 Fs). Multiple regression was used to examine the relative influence of these adolescent variables on the differing parental power dimensions for each parent. Adolescents' sex and religiosity were systematically evidenced as the most important determinant variables. F adolescents consistently perceived the mother as having more power than did Ms, with both sexes similarly viewing the power of the father. Consequently, Fs saw the parental power structure as more equalitarian than did Ms. Religiosity, used as an indicator of traditionalism, was consistently found to be positively associated with parental power perceptions for both sexes, with the exception of parents' outcome-control power (McDonald, 1982) . When daughters raised in second generation polygamous families saw how their mothers got along with each other, how they ran their families, and what accommodations they made to the peculiar demands of the principle in practice, they adapted this learning to their own lives. The important messages that polygamous mothers were inadvertently teaching their daughters were the intricate patterns of relationships—how to live with others in obedience to a difficult principle, how to share both husband and children, and, finally, how to be a female member of a polygamous family (Embry, & Martha, 1985.)

A study of McKeon (1987) deals with understanding the meaning and influence of the sibling relationship in normal development that was then applied to the crisis of pediatric cancer. While the model is designed as a tool to address a vast array of sibling situations, its application to the cancer experience provides a demonstration of its usefulness and a conceptual understanding of the dramatic impact a crisis of this nature has on sibling relationships. Two bodies of literature are extensively reviewed: the research describing the role of the sibling relationship in normal development and the literature relating to the psychosocial adjustment of well siblings of pediatric cancer patients. While the research suggests that siblings can have a powerful and enduring influence on one another and that well siblings of cancer patients are often gravely affected by their siblings' illness, the findings remain primarily descriptive and a theoretical. The purpose of the study was to help organize, within a plausible framework, this growing body of knowledge about sibling behavior.

The ambivalent nature of the sibling relationship is conceptualized as a dialectical tension between the opposing poles of conflict and harmony. In the well-functioning sibling relationship there exists a constant balancing of and interaction between these poles. Rigid, entrenched sibling relationships, in contrast, are marked by exclusively confliction or exclusively harmonious interactions. Parental influence provides the essential "fulcrum" that supports and facilitates the sibling interaction. Salient variables influence the dynamics of the relationship and determine the unique nature of any particular sibling bond. These include sibling contextual variables (e.g., degree of sibling access), parental contextual variables (e.g., parental expectations, parental sibling history), family contextual variables (e.g., coalitions, structure), and socio-cultural variables (e.g., ethnicity, religion).

Expectations about premarital residential independence among young adults were examined based on data from the high School and beyond survey of the class of 1980. The finding that about 70% of the young adults surveyed expects to move out of the parental home before marriage suggests that new norms are emerging that fit into patterns of independence in the transition to adulthood. There is substantial variation in factors affecting expectations about premarital residential independence. Young men more than young women, those with more parental resources, those who expect to marry at older ages, and those who do not have ethnic and religious ties that link them to their parental home until marriage expect to live independently. Religious, racial, and ethnic differences interact in complex ways with gender and SES to influence expectations about premarital residential independence (Goldscheider, Calvin & Goldscheider, 1987)

Further a study was conducted by Smith (1987) in order to know adolescent satisfaction with family life consisting of various antecedents-demographics, family structure, religion, parent/child interactions, adolescent feelings about self, parents' marital and family satisfaction. The sample consisted of 106 families composed of father, mother, and oldest teenager in the home which were drawn from three systematic random samples of approximately 500 families each in Salt Lake, Carbon, and Cache counties in Utah during the last months of 1985. Responses predominantly from the adolescents themselves accounted for 67 percent of the variance in adolescent family satisfaction. Both parent/ child interactions—father support, mother positive control (from mother's perspective), mother negative control, father negative control, and parent-child conflict—as well as adolescent feelings of social worth and self-derogation accounted for a good portion of the variance. It was found that the demographic and family structure variables were poor predictors of adolescent family satisfaction

A study was conducted to examine the perceptions of college students concerning the degree of success with which their fathers performed the role of fathering as well as to identify factors related to perceptions of successful fathering. The DADS Inventory was subjected to a factor analysis which revealed three clusters or underlying factors: communication, commitment, and religiosity. An item analysis indicated that each of the items in the DADS Inventory was significantly discriminating at the.00001 level between those respondents whose total scores fell in the top quartile and those respondents whose total scores fell in the bottom quartile. A reliability analysis indicated Cronbach Alpha values of.96 (communication), .94 (commitment), and and.92 (religiosity). Six major hypotheses were examined and significant relationships were found to exist between the DADS Inventory total scores and the following variables: age of the respondent, race of the respondent, family structure, father's income, educational attainment of the father, depth of religious faith of the respondent, how much the respondent likes his/her father, the degree of closeness the respondent feels to his/her father, the degree of perceived closeness between the respondent's father and mother, the frequency with which the father read to the respondent when a child, the degree of friendship the respondent experienced with the father while growing up, the frequency with which the father played games with the respondent when he/she was a child, the degree of permissiveness/ strictness of discipline which the respondent received from his/her father, whether the respondent received most of his/her discipline from father or mother, the degree to which the father used withdrawal of love, the degree to which the father used reasoning, and the degree of adolescent wellness. These findings add to a growing body of paternal health literature that may enable therapists to deal more effectively with father-child issues and that may serve as a model of paternal success for future fathers (Ray, 1996)

Similarly, a study also examines long-term outcomes of adopted children with special needs, focusing on the identification of family and child characteristics as well as interventions that contribute to family functioning. Drawing on 1995 telephone interview data from 28 families and 69 adoptee located through an adoption program in an urban, southern US city. Analyses suggest that adoptive families with more children, who did not receive family therapy, participated in religious activities, and had fathers with less education, were more likely to have a higher family functioning score (Stephen & Leung, 1998). Perceptions of family strengths of 40 parents and 20 adolescents were investigated by means of semi-structured interviews and Likert-type ratings on commitment, appreciation, spending time together, communication patterns, religious values, and crisis management. Significant differences were found between the perceptions of parents and adolescents for the family characteristics of commitment, time spent together, and crisis management. Although family strength correlated positively with all six identified characteristics, time spent together and appreciation for each other had the highest correlation with family strength. This result supported the notion that sufficient quality time together may be a prerequisite for well-being in family relationships.(Greeff, & le Roux, 1999)

A study of Nicholson, Fergusson, and Horwood (1999) examines the effects of living in a stepfamily during childhood and adolescence on a range of psychosocial outcomes at age 18 years. Data collected during an 18-year longitudinal study were used to examine a sample of 907 children with respect to: exposure to living in a step family during the period from age 6 to 16 years; measures of *psychosocial outcomes* including mental health, antisocial behaviour, substance use, restricted life opportunities, and sexual risk-taking at age 18 years; and measures of prospectively collected confounding factors. The analyses revealed that children exposed to living in a stepfamily for the first time between ages 6-16 years had elevated risks of a range of psychosocial outcomes at 18 years. These included elevated risks of: (1) juvenile offending; (2) nicotine dependence; (3) abuse or dependence on illicit substances; (4) leaving school without qualifications; (5) early onset of sexual activity; and (6) multiple sexual partners. However, these risks were reduced substantially when psychosocial outcomes were adjusted for the confounding effects of antecedent factors such as: *family socioeconomic characteristics*: family history of instability, adversity, and conflict; mother's age, religiosity, and smoking; child gender; and preexisting child conduct and attentional problems. After adjustment, the odds ratios between exposure to a stepfamily and adolescent outcomes were non-significant. Additional analysis revealed that there were no significant differences in outcomes for boys and girls exposed to stepfamilies. It was concluded that although young people exposed to living in a stepfamily had increased risks of poor psychosocial outcomes, much of this association appeared to be spurious, and arose from confounding social, contextual, and individual factors that were present prior to the formation of the stepfamily

Conceptions of parental authority and ratings of parental rules and decision making were examined longitudinally among 82 middle-class African American adolescents and their parents (82 mothers and 52 fathers), who were divided into two groups according to family income. Adolescents were, on average, 13.14 years of age at Time 1 and 15.05 years of age at Time 2. Nearly all adolescents and parents affirmed parents' legitimate authority to regulate (and children's obligation to comply with) rules regarding moral, conventional, prudential, friendship, and multifaceted issues, but they were more equivocal in their judgments regarding personal issues. With age, adolescents increasingly judged personal issues to be beyond the bounds of legitimate parental authority, but judgments differed by family income. Adolescents from upper income families rejected parents' legitimate authority to regulate personal issues more at Time 1 than did adolescents from middle income families, but no differences were found at Time 2. Authority to regulate adolescents' behavior did not extend to other adults or to schools, churches, and the law. With adolescents' increasing age, African American families became less restrictive in regulating prudential, friendship, multifaceted, and personal issues. Adolescents', mothers', and fathers' judgments demonstrated significant continuity over time, but few cross- or within-generation associations in judgments were found. Conceptions of legitimate parental authority at Time 1 were found to predict family rules at Time 2. (Smetana, 2000)

A pilot study explores the degree of solidarity felt between full and half siblings who were raised in a Mormon Fundamentalist polygamous community. The community under study is unique in that, at the level of official culture, it actively promotes full and half sibling solidarity through an ethos that strives to downplay genetic differences in favor of a harmonious family living together in one household.

This community is an ideal cultural setting in which to examine the suitability of inclusive fitness theory for understanding the factors that promote family cohesion, sibling solidarity, and rivalry. The main question such as: is the degree of sibling solidarity a manifestation of genetic closeness or a natural byproduct of emotional closeness that arises from being raised together? Result revealed more solidarity between full siblings than between half siblings. Our data suggest that, despite the force of religious ideals, and notwithstanding the continued close physical proximity of half siblings in the polygamous family, there was a pronounced clustering of feeling and affection in the polygamous family that was found consistent with inclusive fitness theory (Jankowiak & Diderich, 2000).

Article of Semans, and Fish (2000) highlights findings from a qualitative analysis of the ways in which Jewish families identify how Judaism influences their lives. A theoretical sample of two religious and two cultural families were chosen from a larger sample of 48 Jewish families in Central New York. The qualitative part of this study, which was part of a larger multi-method investigation, was done in order to gather inductively any data that would allow the researchers to build a theory about a particular type of ethnic identity—Jewish identity—and how it affects family dynamics. Eleven categories emerged from this study suggest that particular type of ethnic identity influences many family dynamics, for example, styles of communicating, parenting, values, and family rituals, played important role. The participants seem to dissect the world with a "Jewish scalpel." That informs their daily interactions, parenting styles, and their self-perceptios. Recent research indicates that there are significant changes occurring in this normally static group. In particular, a shift from community to family as the center of Hutterite life and decision making was noted. This appears to be mainly due to greater affection and equality in marriage, as well as greater societal influence in dating and fertility practices. Further, each area of the family life cycle- kinship, early childhood, the school years, adolescence, marriage, fertility, and later life may leads to deeper changes is well-being (Ingoldsby, 2001)

A study of Steinberg (2001.) examines the most important ideas to have emerged from the last 25 years of research on adolescent development in the family context and suggests some directions for the future. Two major sets of questions organize the review. First, how can we best characterize normative family relationships during adolescence, and, more specifically, is adolescence a time of parent-child conflict? Second, how do variations in parent- child relationships affect the developing adolescent? The answer to the first question depends on what is meant by conflict and, more importantly, from whom one gathers data. There is a need for a new perspective on the family, one that emphasizes the different viewpoints and stakes that parents and adolescents bring to their relationship with each other. Special attention should be paid to studies of the mental health of parents of adolescents. With regard to the second question, it is argued that there is enough evidence to conclude that adolescents benefit from having parents who are authoritative: warm, firm, and accepting of their needs for psychological autonomy. Therefore, it would seem most beneficial to institute a systematic, large-scale, multifaceted, and ongoing public health campaign to educate parents about adolescence, one that draws on the collective resources and expertise of health-care professionals, scientists, governmental agencies, community organizations, schools, religious institutions, and the mass media

The studies of **Elbedour, Bart, and Hektner (2007) deals with 210** adolescents randomly selected from Bedouin Arab (mean age 15.9), who were administered instruments assessing their family environment and mental health. The second study involved 182 Bedouin Arab adolescents in which the student participants completed a single instrument about themselves and in which the teachers of the students completed the Teachers' Report Form of the Child Behaviour Checklist. The Bedouin Arab adolescents fell into two groups: (a) adolescents in monogamous family structures and (b) adolescents in polygamous family structures. The findings of the first study suggest that the two groups did not differ significantly in the majority of the assessed variables, even though there were significant differences obtained between groups for 4 of the 13 assessed variables. The two groups did not differ significantly in the second study.

The important analysis that cxamines the direct and indirect influences of early gender socialization on the allocation of routine housework later in the life course was studed deeply. The study articulates hypotheses that the relationship between gender and socialization early in adulthood and housework allocation later in adulthood is modcrated by gender and union type and is mediated by subsequent contextual characteristics of the couple. The analysis draws on panal data from a sample of 586 young adults spanning 31 years. Finding indicate that married men's attitudes about gender early in adulthood were more influential for the later division of labour than are married women's attitude, but gender differences in the influence of early attitudes on later housework patterns are not present among cohabiters. The influence of early gender socialization on later housework allocation is mediated by couple-level resources and time availability among cohabitoes but not among married individuals.-Journal abstract (Mick, 2005)

In a classic study effects of polyandry versus monogamy were assessed for the beetle Phoracantha semipunctata (Coleoptere: Cerambycldae) by comparing measures of females reproductive success, include fecundity, egg viability, time until exclusion, and clutch size. The effect of intermittent presence or absence of a male was also evaluated. Polyandry was detrimental to female reproductive success in comparison to monogamy. Fecundity, egg viability, and clutch size were lower and time to elusion were increased for eggs from females with multiple mates compared with monogamous females, intermittent presence of males had no effect on females reproductive success. Possible explanations for the decreased fecundity experienced by females with multiple mates included sperm competition intensity, costs of male harassment, and exceeding the optimal mating frequency. Females may reduce costs associated with polyandry by spending less time on host logs where mating occurs (Bybee, Linda, Millar, Jocelyn, Paine, Timothy, Campbell, & Kathleen et al, 2005)

Similarly a study aimed at investigating and comparing general health and life satisfaction of students in polygamy and monogamy families. The sample size of the research consist of 128 students (32 male and 32 female from polygamy families and 32 male and 32 female from monogamy families) were selected randomly. To collect the data General Health Questionnaire and life satisfaction Scale of Student were applied. The result of the present study showed that there was negative significant correlation between physical symptoms, anxiety, social dysfunction and depression sub-scales and total scores of GHQ and life satisfaction, and also the result of the ANOVA indicates that there was non-significant difference between the mean scores of four groups on general health scale and sub-scale. The result did not emerge significance gender difference (Tamini & Kahrazei, 2010).

3.3 TRIBAL LIFE VULNERABILITIES AND RESILIENCE

Since the tribal community all over India have been subjected to various forms of deprivation such as, alienation from land and other forest resources since the British rule, as a part of tribal society, women have to suffer and yet survival for the sake of their children and her community. In the context of the India's plural social structure there have been many autogenous movements against colonial imposition, many of them ultimately converging into a single anti-colonial national liberation struggle. In the 1981 census out of India's total population of 68, 51, 84,692, as many as 51, 86, 28,638 were enumerated as belonging to the scheduled tribe communities. Out of the latter 2, 60, 38,536 were males and 2, 55, 90,103 were females. Thus among the scheduled tribes, per 1,000 males, the number of females were 983. This compared favorably with the total population and scheduled caste population, with sex ratios of 934 and 932 respectively. Higher sex ratios among tribal in all the states and UTs except three might be the outcome of several factors, to wit, differentials in sex-linked birth rate, neonatal and infant mortality, maternity death and survival at higher age.

Vidyarthi has classified the tribal communities as (a) forest hunting type, (b) hill cultivation type, (c) plains agricultural type, and (d) simple artisan type. Roy Burman elaborated the classification to cover (a) forestry and food gathering, (b) shifting cultivation, (c) setting agriculture, (d) agricultural labour, (e) animal husbandry, (f) household industry, and (g) miscellaneous occupations including industrial employment. Before the discovery of agriculture, for thousands of centuries, man lived on hunting, fishing and food gathering. The early division of labour was on the basis of age and sex. Males would engage themselves in physically arduous tasks of hunting and manufacturing the heavy stone implements for hunting. The females would more frequently collect roots, tubers and edible leaves, prepare the food for the family or the horde.

Fig: 3.3.1 (a): kinnauri women in a light mood while attending marriage

A tribal woman occupies an important place in the socio-economic and political structure of the society. They exercise free and firm hand in all aspects related to their social and economic life. But it is still important to emphasize that the tribal woman is in herself exactly the same as any other woman with the same passion, love and fear, the same devotion to the home, to husband and to children, the same faults and the same virtues. Though the tribal women are away from the main stream of nation life, but they are not kept away from the main stream of national life, but they are not kept away from the impact of socio-economic changes affecting the neighborhood society in general. In the process of change, the tribal woman is forced to adhere to certain norms which may even take away her freedom, her control over the traditional productive system, her house, family and children and even her own life. The process of such alienation has an impact on the tribal women.

Age at marriage is considered to be an important factor affecting the birth rate. The 1971 census data show that whereas among the total population 11.76 per cent of the females got married before the age of 14 years, among the scheduled tribes the corresponding figure was only 2.6 per cent. In 1971, 7.4 per cent of tribal women of all age groups were widows and 0.50 per cent was divorcees or separated. Among most of the tribes all over India, customary payments are required to be made by the bridegroom to the bride's father during marriage. Sometime, the amount is so high that the prospective groom has to work for years, so as to save enough to be able to marry. For a girl to wait till her mid 30s, before she can find someone who can pay and become her partner in life. Sometime elderly males acquire more than one wife by paying the bride-price. The participation ratcs of the male and female working force among the tribe's are much higher than that among others. Participating rates of both male and female workers among the scheduled tribes were higher than those of both scheduled castes and total population. Ethnographic literature shows that non-participation in the working force in the primary sector of economy is an indicator of social status.

It was this division of labour that led to the first technological revolution. It provided an opportunity at the primitive women to observe the behaviour of the plants in different situations. It could not be very often that she unwittingly left the remnants of wild edible grains on the soil and some time or other she would become aware that from these grains sprouted and grew plants to yield more grains. This knowledge, however, created an aura of luminosity in the darkness. The woman, who was the better half of man, was transformed into the mother-goddess. Her mysterious essence rose from the travails of mundane existence. But in the process her physical self was left behind to wallow in the day-to-day life of deprivation and subjugation. Collection and gathering of forest products constituted important source of livelihood not only among the non-agricultural tribes, but also among many agricultural tribes.

The evolved form of slash and burn cultivation augmented the supply of food, but until recently in the primitive state of sanitation and hygiene, there was no corresponding rise in population. This allowed the man to resign from hunting to a certain extent. It is not surprising therefore that males often tend to live a parasitic life on the labour of the females, and while away their time in smoking narcotics or drinking alcohol in the societies practicing slash and burn cultivation. Man was more associated with animals in the operations of hunting and herding. Plough culture with the aid of draught animals by earlier association, his job increases the quantum of work in sowing, transplantation, weeding, harvesting, transport and threshing.

Fig 3.3.2 (b): Women making chapatti to baratis (Janya) in marriage

Erosion of women's role due to cultural is found in the non-agricultural sector as well as has been seen, fraternal polyandry is a form of marriage that was prevalent among the *tribal* class. Traditionally, marriages were arranged by the parents, often when the children were still very young. The marriages were decided for patrimonial reasons; the brides' and bridegrooms' personal preferences were of no consequence. In polyandrous conjugal family, the eldest brother is more often than not, the dominant person in the household. All the other brothers, however, shared equally the work, and had the right to sexual relations with their common wife, who had to treat them equally. All children were treated equally, and a "father" is not allowed to show any favoritism, even if he knew who his biological children really were, as biological paternity was not regarded as important. Similarly, the children considered all their uncles as their fathers, and a child avoided treating members of the elder generation differently, even if they knew who their biological father was. Divorce was quite simple. If one of the brothers in a polyandrous marriage felt displeased, he only had to leave the household. Polyandrous marriages were often characterized by tensions and clashes for a variety of different reasons. Conflicts may, for example arise because a younger brother wanted to contest the authority of his eldest brother, or sometimes, sexual favoritism might occur, generating tension among the male partners in the marriage, especially so, when there are significant age differences among the brothers.

The relationship between level of piety and level of violence between siblings was researched using a secondary analysis of survey data gathered by Straus on a sample of 345 unmarried students reporting on their home situations in their last year of high school. Piety, the independent variable, encompasses three aspects: the associational, orthodox, and devotional (Lenski, 1961: Stark & Glock, 1968). It is operationalized in terms of an index calculated on the basis of answers to questions regarding: (1) the frequency of church attendance and (2) the frequency with which one consults God when faced with difficult decisions. Sibling violence, was, is measured by responses to questions regarding the use of violent acts such as kicking, pushing, throwing an object, etc, to resolve conflicts with a sibling. SES, the sex of the R, and husband-wife dominance were used as control variables. Examined are the effects of: the R's piety upon level of violence with a sibling; mother's and father's piety upon sibling violence; and level of family piety homogeneity upon sibling violence. Findings suggest that in pious families where father's dominance is less than or equal to mother's, sons tend to be above average in violence toward a sibling (Carol, 1981). This project proposes to combat the rising rate of violent crime and social disintegration among young Black males by preparing Christian men to serve as mentors to neighborhood youth. The project conducted three seminars for church leaders concerning child abuse and its effects, self-esteem among Black males, and substance abuse. These mentors were paired with seven first offenders referred by the juvenile court. Four of these young men significantly improved in academic performance, citizenship behavior, and family relationships (Ramey, 1996)

Violence has reached epidemic proportions in the United States, with African American males residing in urban areas bearing the brunt of this epidemic. The violence permeating our society emanates from a variety of societal ills, including poverty, racism, substance abuse and exposure to violence. Traditionally, methods of research on adolescent violence have focused on an identification of associated risk factors. The majority of African American adolescents living in communities with widespread and chronic violence grow up to be law-abiding citizens. The reason for this may be due to resiliency within these adolescents. Resiliency has been defined as the ability to experience adverse circumstances and successfully overcome them. Resiliency is enhanced through three mechanisms identified as protective factors: individual characteristics/traits; familial traits; and extra familial relationships. The risks faced may be altered, and even ameliorated, through the presence of these protcctive factors. The purpose of this paper is to identify protective factors within the family that foster resiliency. Protective factors within they family may be categorized as having three broad characteristics: caring and support, high expectations, and encouragement of participation and involvement for the adolescent (Calvert, 1997)

Loneliness is a prevalent and serious problem among adolescents. Changing family structure and characteristics of the parent-child relationship influence attachment sentiments of adolescents predisposing them to the experience of loneliness. In the present study, parent-child relationship was analyzed as a predictor of adolescent loneliness. A theoretical model provided the framework for the individual and multi-factorial analyses of parent-child relationship factors and family structure on loneliness prediction. A subset of the National Longitudinal Study of Adolescent Health (Add Health) provided the study sample of 5,201 with 1,750 adolescents reporting loneliness. Lonely adolescents were more likely to be female and older than adolescents who were not lonely. Black adolescents were lonelier than White non-Hispanic or Hispanic adolescents. Adolescents with high religiosity scores were less likely to be lonely than adolescents with lower scores. Using logistic regression analyses, it was found that parent-adolescent connectedness, parent-adolescent presence, parent-adolescent activities, closeness to the non-residential parent, and family structure are significant predictors of adolescent loneliness. Lonely adolescents were more likely to be from one-parent or blended families, feel less connected to residential and non-residential parents, have few opportunities during the day to be with parents, and engage in few parental activities (Antognoli-Toland, Lucia, 2000).

Adolescent substance use has become a serious concern nationwide. Although there are many ways of viewing adolescent substance use, family influence has been established as one of the strongest sources of risk and protection. A review of the literature indicated relevant theories for understanding adolescent substance use and specific family-based variables influencing adolescent substance use. In general, there seems to be a relationship between adolescent substance use and family-based risk and protective factors. Relevant theories identified in the literature review include family systems theory social cognitive theory, social control theory, and strain theory. Specific family-based risk and protective factors include family relationships such as with siblings and parents and family characteristics such as ethnicity and religious backgrounds. Future implications for research and prevention/intervention in relation to family-based risk and protective factors are discussed (Vakalahi, 2001)

3.4 PSYCHOLOGICAL INTERVENTION AND WELLBEING

The psychological interventions since time immemorial have proved useful for promoting wellbeing. In a study (Glen, 1962) identified seven types of social structures in the child rearing relationship. They were autocratic, authoritarian, democratic, equalitarian, permissive, laissez-faire, and ignoring, those affect the wellness. The investigation was concerned with two problems: (1) the analysis of the relationship between social class, education and religion of parents and size of family, and the child rearing structures, and (2) the evaluation of structural effects upon the affective relations between parent and adolescent, and the adolescent's attitude toward parental child rearing policy. This study was based on 7,400 adolescents residing in Ohio and North Carolina; the data were obtained with a structured questionnaire. Parental dominance was most common to parents of low socioeconomic status, who are Catholic, and who have large families. The likelihood of mutual rejection in parent-adolescent relations and unfavorable evaluations of parental policy was greatest in autocratically structured relationships.

The similarities and differences in family stress, availability of resources, and levels of coping as reported by adolescents in nuclear, single parent, and remarried families has been examined in this study. The 1,277 respondents were selected from a larger sample who participated in a statewide project funded by the Florida Department of Education. Four instruments were administered to Home Economics Family Living students by their teachers: (1) a background information questionnaire; (2) the Adolescent-Family Inventory of Life Events and Changes; (3) the Family Inventory of Resources for Management; and (4) the Adolescent-Coping Orientation for Problem Experiences. Adolescents' scores indicated that members of nuclear families had experienced fewer family life events or changes and less family stress than adolescents in single parent and remarried families. The level of available resources was higher for adolescents in nuclear families than for those in the other family forms. Despite differences in the number of life events or changes, levels of family stress, and availability of family resources, the coping scores among adolescents were not significantly different across family types. Sex of the adolescent was identified as an important influence on stress levels in nuclear and remarried families. Adolescents' coping scores were most influenced by the sex of the adolescents in all three family forms and by the importance of religion in nuclear families. Social class appeared to influence perceptions of available resources in nuclear families but appeared to have little effect in other family forms. Social class and the importance of religion appeared to be predictive of the level of available resources in nuclear families. In all three family types, a higher correlation was found between students' coping scores and their resource scores than between their coping scores and perceived level of family stress (Stillwell, & Taylor, 1988).

Further, a study investigated and compared the experiences of paternal nurturance by early adolescent children. It was designed to determine if there was a difference in a father's nurturance as experienced by the early adolescent child in varying groups in the select categories of sex, ethnicity, religious affiliation, family configuration, and family income. The child's experience of paternal nurturance was measured with the Buri, Misukanis, and Mueller Paternal Nurturance Scale. This is a 24-item Likert-type scale used to measure parental nurturance from the viewpoint of an adolescent evaluating the nurturance he or she has received from his or her father (Buri, Hengel, Komar, & Richtmeier, 1989). A child begins acculturation within the setting of his/her family. There is a commonness in the socialization of persons experiencing the same culture and society. Parents and family members bring this into the home of the child; however, mothers and fathers each represent differing modes of socialization (Johnson, This study focused on the early adolescent child's experience of a father's nurturance. Aspects of the father's nurturant activity affect attachments, self-concepts, self-esteem, gender orientation, and cognitive development. Normal and healthy development in all of these areas assist socialization of the child in preparation for a wholesome and satisfying life in the adult world. This study tested twenty Null Hypotheses of a child's experience of paternal nurturance by the father. Ten of the hypotheses were supported and ten were rejected. When one compares variables of ethnicity, family configurations, religion, age, and income, it shows that there are differences in experienced nurturance among children within these groupings. The findings of this study provide information for parents, religious leaders, teachers, counselors, and persons working in the delivery of social services. Information in this study is also of value for family life awareness programs. It can alert fathers to their parenting habits, the needs of their children, and with this awareness, possibly help them avoid future family crises (Schoellmann, 1990.)

Using longitudinal design this research assessed the degree to which an adequate conceptualization and measurement of religious family values could be conducted. Questionnaire responses from 560 young adult males, originally studies in 1981, were analyzed revealing a cluster of values centering around family home evening, scripture study, family prayer, and moral behavior formed one dimension of religious family values. Three other related value dimensions were also identified, viz. birth control, divorce, working mother. Result showed that the family, peer, and religious influences all contributed to explaining variation in young adult religious family values. Adolescent religiosity emerged as an important intervening variable which also influenced whether or not the young man chose to serve a mission for the church. Of the exogenous variables, home religious observance was the single most important influence on young adult religious family values. The direct effect over nine years suggests the strength of family socialization in a specific relationship to special values. (Orvel, 1992.)

Teen marriage may be a way of legitimately culminating a sexual relationship and attaining adult status. The purpose was to investigate whether the religion in which a young person was raised has an impact on the decision to marry early. Extrapolating from our findings on premarital sex as well as previous research of others, we hypothesized that

Fundamentalist and Institutional Sect backgrounds produce higher rates of teen marriage. The data between 1979 and 1984 for Whites and female Blacks was collected. The logistic regressions indicated substantial differences in the likelihood of teen marriage by religious heritage category for male and female Whites, but not for female African-Americans. Using mainline Protestants as the comparison group, we find that young Whites with Fundamentalist and Sect-like backgrounds were much more likely to marry by age 19, while Catholics and non-Christians are significantly less likely to marry early. These differences persist even when controlling for geographic factors, parental and family characteristics, church attendance, and expectations for adult roles 9 (Hammond, Cole & Beck, 1993)

A family process model of Brody, Flor, and McCrary (1994) links maternal and paternal formal religiosity to marital interaction quality, parental co-caregiver support and conflict, parent-youth relationship quality, and parental use of inconsistent/nattering parenting strategies. The sample included 90 African American youths and their married parents, who lived in the rural South. African-American community members participated in the development of the self-report instruments and observational research methods used to test the model. The results supported most of the hypotheses. Religiosity was linked with higher levels of marital interaction quality and co-caregiver support, and with lower levels of marital and co-caregiver conflict. The associations between religiosity and parent-youth relationship quality were mediated by the marital and co-caregiver relationships The relationship between a parent and a child is crucial to the child's development. This study addresses two areas of adolescent parenting in which there exists a paucity of research:

(a) teen mothers who are successful in developing a supportive relationship with their children versus those who are having difficulty and (b) the intergenerational family influences on the mother-child relationships of these teen mothers.

This qualitative descriptive study used a synthesis of styles and methodologies to explore the similarities and differences in the intergenerational families of teens who have a supportive relationship with their babies and those who do not. A standardized mother-child interactional instrument was used to score the quality of mother-child videotaped interactions between 106 mothers and children participating in a mid-Michigan adolescent parent program. Five participant families (a teen and one adult from each family) were then selected from each of the score distribution extremes to participate in an interview. In the interview, each family told their family story and constructed a family genogram. A constant comparative analysis of the interview transcriptions and genogram identified three core categories: family frame, family characteristics, and family function. Differences between the two groups of family participants pointed out several persistent intergenerational stressors that appear to influence the ecological system of these families: family size, family spacing, generational compression, cultural origin and mobility, socioeconomics, housing, religion, relationships, perceptions of family status, role of the rule makers and breakers, and daily routines. These social stressors experienced by several generations seem indicative of intergenerational distress. The greater the amount of intergenerational distress expressed in the family stories of the informants, the more difficulty the teen appeared to have in forming a supportive relationship with her baby (Mary, 1994)

A study of Keysar, and Kosmin (1995) demonstrates that religion is significantly associated with the acquisition of postsecondary education by white women in the contemporary United States. Religion has both direct and indirect effects on educational attainment. Religious traditions differ in the degree to which they emphasize the importance of the family, marriage, and child bearing. This, in turn, influences how much higher education the women of the group are likely to obtain. Thus, religion has an indirect effect on the educational levels of women through their demographic behavior. In addition, we show that there is a relationship between religion and the education of white women that is maintained beyond other socio-demographic factors. A refined model involving 12 religious identifications on a conservative-liberal continuum, subjected to multivariate analyses, illustrates that educational differences tend to be wider among older women. Surprisingly, conservative protestant and no religion adherents do not form the polarities, but have similar middle-order levels of educational attainment

In the troubled economy of the rural Midwest, adolescents come of age with a future that seems increasingly less promising at home than in other places Their growing recognition of limited opportunities may fuel a resolve to migrate to other regions, which clashes with binding attachments to local people and places. By using preferences for living near family and in the local community, obtained in the 8th and 11th grades, this longitudinal study modeled the social and developmental pathways by which adolescents approach decisions to leave home and settle in other parts of the country. Data come from 351 two-parent families in the Iowa Youth and Family Project, launched in 1989 to investigate the economic stresses and family consequences of the farm crisis. Lack of socioeconomic opportunity, relatively weak and declining ties to parents, kin, and the religious community, and strong educational prospects emerged as potent sources of a declining preference for Living near family and in the local community among boys and girls. Whether coupled with family attachments or not, plans to settle elsewhere after education are linked to more elevated feelings of depression and unhappiness about life (Elder, King, & Conger, 1996).

A descriptive exploratory design was used to examine the relationship of socio-demographic characteristics, family functioning levels, family and community socioeconomic status, and peer relationships to adolescent sexual behavior. Four questionnaires (Demographic Profile, Family Adaptability and Cohesion Scale, Parent-Adolescent Communication Scale, and an Index of Peer Relations) were administered to 101 African American female adolescent volunteers. Means, standard deviations, and frequency distributions were used to describe sociodemographic characteristics of the subjects, and frequency distributions were used to describe the sexual activity of the subjects. The associations between the dependent and independent variables were examined, using the Chi-Square statistic, Fisher's Exact test, and t-tests. Results indicated an association between age, neighborhood socioeconomic status, church attendance, grade point average, educational aspirations, and family cohesion levels and sexual activity status among the subjects. Although the differences fell short of statistical significance, trends related to parent's home ownership, family types, parent-adolescent communication, and single-parent household structures were consistent with those of earlier investigations. Implications for nursing include assessing family functioning and the interrelationships of family members and the adolescent in the family setting. The results of this assessment can be used to guide clinical programs for counseling and strengthening families with adolescents (August-Prudhomme, 1997)

A study conducted on community-based program designed to delay onset and reduce the frequency of alcohol and other drug (AOD) use among high-risk youths, ages 12 to 14, through strengthening family resilience. It is part of a larger five-year demonstration project funded by the Center for Substance Abuse Prevention (CSAP). The program was implemented in multiple church communities in rural, suburban, and inner-city settings. Program components of this study included parent or guardian and youth training, early intervention services, and follow-up case management services. The results show that the program produced positive direct effects on family resilience. The evaluation also found positive moderating effects on delayed onset of alcohol and other drug use and frequency of alcohol and other drug use among youths in the form of conditional relationships with changes in those family resilience factors that were targeted by the program (Johnson, Bryant, Collins, Noe, Strader & Berbaum, 1998)

This study of (Hughes, 1999) explores how various aspects of a grandparent's religiousness are related to involvement with their grandchildren. The study ascertains whether religious grandparents are significantly more involved with their grandchildren than less religious grandparents by testing a series of bivariate and multivariate regression models. Multiple measures both of religiousness and of grandparent involvement are used. The study also examines factors that potentially confound, link, or explain the connection between religious involvement and grand parenting (e.g., demographic factors, health, community involvement, traditional values, ties to others). The sample of 585 grandparents comes from two related studies of rural families, the Iowa Youth and Families Project and the Iowa Single Parent Project. Overall, the results indicate that religious grandparents are more involved grandparents, and this involvement is explained in part by their generally greater involvement in all types of family and social ties-religious grandparents are more likely to be enmeshed in social ties to others. The implications of these findings are discussed in light of prior research that finds little influence of religion. We also consider the implications of our results for the future of grandparent involvement

The use of religion in the lives of adolescents to repair problematic or disrupted attachments is discussed in the context of attachment theory and Kohut's self-psychology theory, with particular reference to the self-object. It is proposed that adolescents do not seek to break ties with parents or adults so much as to revise their relationships in a more adult direction. Two adolescent cases, one beginning treatment in childhood and the other in early adolescence, are presented and discussed in the context of attachment theory and self-psychology (Lovinger, Miller, & Lovinger, 1999). Using data taken from the second wave of the 1992-1994 National Survey of Families and Households, the influence of religious affiliation, attendance, and ideology on father's involvement in one-one-one activities, dinner attendance, nonreligious youth-related activities, and religious youth groups is examined. Findings reveal religious effects for each of these four measures. Moreover, the competing hypothesis is tested, ie, that these effects are artifacts of a conventional masculinity such that the type of men who are more conventional (in work, civic activities, and gender role ideology) are, as a consequence, both more religious and more involved with their children. No support for this competing hypothesis was found, which suggests that religion has an independent effect on paternal involvement. Morever, it was revealed that two indicators of this conventional masculinity - job satisfaction & civic engagement - are positively associated with paternal involvement (Wilcox, 1999).

Many interventions have addressed risk-taking behavior as they pertain to sexual risk behaviors. Such interventions have mainly addressed the influence of peers and the environment in adolescent risk-taking behavior. Few studies have addressed the role of family environment in adolescent risk-taking behavior. The aims of the study were to develop a measure to assess family environment, to evaluate the psychometric properties of the family environment measure, and to assess associations between family environment constructs and outcome variables, such as condom use, sexually transmitted diseases history, drug use history, and parental communication. The study assessed 4 constructs of family environment: family cohesion, family conflict, family expressiveness, and family moral religious emphasis. The findings from this study showed that family cohesion was a significant predictor of healthy behaviors among African American adolescent females 14-18 years old residing in Jefferson County, Alabama (Parker, 2000)

The purpose of this research was to determine Turkish women's satisfaction with their marriage and to establish a standardized scale of marital satisfaction to use for this purpose. This research used descriptive and methodological designs. The population for this study was a sample of Turkish women who were married, and were applying to five primary health care centers in Erzurum, Turkey. Four hundred and ninty-five women were selected by a simple random sampling strategy from a larger population of women, Data were collected using a marital satisfaction scale between April 1 and June 30 of 2002. The Cronbach's alpha for this scale was 0.96. factor analysis was conducted for the 15 items and a one factor solution emerged that was designated the marital satisfaction scale. The mean score of the scale was 78.54 and the mean score for the items was 5.2 points, which revealed that the women were somewhat satisfied with their marriage. The result of this study cannot be generalized to all the Turkish women because the sample represented only one area of Turkey. The validity and reliability of the scale were supported as a measure of marital satisfaction that could be helpful for use in health care centers and offices (Erci, Behice & Raziye, 2005)

This article considers some of the common issues with which marital couples in India struggle. These issues arise from the way that Indian families are typically struggle and the rules that govern family relation ship. Ideas from the social constructionist and contextual approaches are suggested as helpful interventions (Sonpar, shobna, 2005). Finally a study explores the process of relationship development in marriage that foster personal and relationship growth, semi-marriage that promote individual and relationship growth. Grounded theory methodology was employed and 'Empowering Connections" emerged as the core themes. Empowering connections are characterized by a process of growth captured by the following categories: "Connecting Emotionally," "Loving respectfully," "Expanding One's Self," and "Experiencing Empowerment Connections." The result of the study contributed to the understanding of marriage and provided insight to premarital couples, people who want to improve their marriage, and to therapists working with couples (Hurst & Nancy, 2005).

QUALITATIVE METHOD

It is impracticable not to be flabbergast by India where civilization present itself in such a dizzying and innovative burst of cultures, religions, races and diverse tongues, intertwining in a thread of unity despite diversities. Every aspect of the motherland presents itself on a substantial, exaggerated scale, worthy in comparison only to the superlative mountains that overshadow it. Perhaps the only thing more difficult than to bc indifferent to India would be to figure out, foresee and be in command of its people burgeoning under the Himalayas, touching three large bodies of water thereof form mammoth and terrestrial beak imaginable in its nature and functioning. The Himalayan region plays a vital role in determining India's natural resources base, eco-system, climatic conditions, river system and strategic and security parameters. It roughly assists 12 per cent of the geographical areas and 6 per cent of the total geographic areas and 6 per cent of the country's population. This region includes Jammu and Kashmir, Himachal Pradesh, U.P. & Uttrakhand Hills, Sikkim, Darjeeling and entire North East. The region varies in terms of cultural pattern, ethnic composition, level of socio-economic progress, resource base, human settlement and technology.

The (H.P) state is proclaimed by ancient Indian texts as Devbhumi — "Land of the gods and goddess" — and believed to be the earthly home of the mighty Lord Shiva. This beautiful, far-flung region has an almost palpable presence of divinity. Etymologically, Hima in Sanskrit means snow Anchal means place or snow-bounded place named by one of the great Sanskrit scholars of Himachal Pradesh, Acharya Diwakar Datt Sharma. After independence the Chief Commissioner's Province of H.P. came into being on 15 April 1948. The province comprised the hill districts around Shimla and southern hill areas of the former Punjab region. Himachal became a part C state on 26 January 1950 with the implementation of the Constitution of India. Himachal Pradesh became a Union Territory on 1 November 1956. On 18 December 1970 the State of Himachal Pradesh Act was passed by Parliament and the new state came into being on 25 January 1971.

Presently the state is has spread up in an area of 55,673 square Kilometers divided into the twelve districts followed with 17,495 inhabited rural villages(90%). Hamirpur district is the smallest of all whereas the Lahaul and Spiti is the largest. According to Census 2001 there are 60, 77,900, population here with a density of 109 persons. Out of the total population, the number of males and females ratio is 30, 87,940 and 29, 89,960 respectively. This state has the utmost percentage of rural population among all the state of the country. The state is performing well in the education sector where the literacy rate is 76.45% (85% for male + 67.40% female). The history Himachal Pradesh dates back to the time when the Indus valley civilization flourished between 2250 and 1750 BC. The Aryan influence in the region; Anglo Gorkha War, and the British colonial government made the state as quite memorable. Tribes such as the Koilis, Halis, Dagis, Dhaugris, Dasa, Khasas, Kinnars and Kirats inhabited the region from pre-historic era. During the Vedic period, several small republics known as "Janapada" existed which were later conquered by the Gupta Empire. After a brief period of supremacy by king Harshavardhana, the region was once again divided into several local powers headed by chieftains, including some Rajput principalities. These kingdoms initially enjoyed a large degree of independence later were devastated by Muslim invaders number of times. Mahmud Ghaznavi conquered Kangra at the beginning of the 10th century. Timur and Sikander Lodi also marched through the lower hills of the state and captured a number of forts and fought many battles. Several hill states acknowledged Mughal sovereignty and paid regular tribute to the Mughals. The Gorkhas, a martial tribe came to power in Nepal in the year 1768. Gradually the Gorkhas annexed Sirmour, Shimla and Kangra. With the leadership of Amar Singh Thapa.

Kinnaur district is known with various names such as Kinnaura, Kinara, Kanwara, and Kannaura. There are many tribes and caste in India which claim their origin from divine source and the "Kinners" is one of them that belong to such a divine source. These people are described as distinct race of demi-gods, somewhere between human being and gods. They are described in numerous ways in the literature. For example, the famous Sanskrit classic "Amarkosa" reckon it Gandharvas, Yakshas, Vidya-dhavasm, Apsaras, Siddhas and Guhayakas. Vishnupuranas has described them as a half man and half a bird. Still other description depicts them with a human body and horses head. According to "Bhagvad Purana" Kinner trace their origin from the shadow of Lord Brahma. The Kinnauras are generally the Scheduled Tribe. Entire elite community of the country is in the stereotype that in this district there do exists only tribes of high castes such as Kanet, Khas, and Khoshia who are the dominating group. There is no defining the fact that there also do exist caste hierarchies in the district. There also do survive Chamang (Koli), Domang (Lohar) Badhi (Mistri) and Nagalu (basket makers) here. The people belonging to the lower caste handle the work of carpenter weaving, artisans, ploughmen (hali), goldsmith, silversmith, carpenters, masons as well as beating Dhol (drums) and Nagaras for pleasing elite people as well as their Devi and Devta. It seems that the excess religious beliefs of the tribals in their Devi and Devta are one of the reasons for their social discrimination, untouchability, exploitation as well as poor cognitive abilities and backwardness. But the dominating group surmounts pressure though local deities and threat them to serve them as status qua for pleasing their Devi and Devta. It perhaps becomes a greater barrier for their personality development. Another factor is the poverty and prolonged deprivation that propel them to do such a menial job to survive their families and offspring. Beside this tough geographical region and excessive show-fall fuether make people on vulnerable toward limited knowledge, rudimentary customs and rites.

4.1 PRESENT STUDY

The present study is confined to district Kinnaur, which is approximately 270 Km from Shimla town. The study is pioneer in nature because attempt has been made to explore the psychological wellbeing among polyandrous and monogamous socially disadvantaged tribal people of Kinnaur. The social institution and their racial or caste system have not been studied so far in a more comprehensive way by the social scientists. So far, it is generally believed that there exists only tribal population in tribal areas whereas the scheduled caste has remained dormant in the eyes of the researchers especially in academician, policy makers and politician in the contemporary scenarios. The reality is that there do exist complex socio-cultural heterogeneity among the hierarchies of castes exists in this rural tribal district. One of such system is quite visible through caste-system practices as well as in marriage institutions.

Indians regard the institution of marriage as a sacrament. For Hindus, it"s a commitment and strong social bond lasts forever between a man and a woman. Therefore is divine union of two soul, after marriage. But the contemporary way of marriages and weddings in India is very different from what it was years ago. But the trend is entirely different and conservative in ancient times. There are various types of marriages in India in general as well as in Himachal Pradesh and Kinnaur in particular. The Brahma Marriage prevailed gurukul system, where the boy use to go to live with his guru to attain worldly knowledge and expertise and after completion of his studies he marries to a girl *thereafter it is called Brahma Marriage.* Another form is Daiva marriage where the girl is married to a priest during a sacrifice. Further in Arsha marriage the daughter bond is made with the saga. In Prajapatya marriage father gives his daughter to the bridegroom, by addressing the couple with the mantras by following dharma principle. The Gandharva Marriage or Gandharva Vivaha is a celestial's type simply involves exchange of garlands after which the marriage is confirmed to the bridegroom. In Asura Marriage, an unsuitable bridegroom approaches the family of the bride and offers a high price to get married to the girl. In Rakshasa marriage the groom fights battles with the bride's family, overcomes them and takes her away from her family and then forces her to marry. The Paishacha marriage is usually consider as the most inferior type of marriage. In this marriage, no one considers the girl's wishes, whether she wants to marry or not instead. She is just forced to marry.

In tribal areas there is a polygamous marital system beside monogamy. The polygamous people practices polyandry practices both in upper and lower caste groups in district Kinnaur. The polyandry cases are rare and are consider harmful to their culture. But the polyandrous groups share one wife among real brother who is known as fraternal polyandry as well as in other relative known as non-fraternal polyandry. It seems that the female, in this system is like commodities are purchased and sold by giving appropriate bride price. But the Kinnauri women are very hard-worker multifariously such as in the fields, cut and harvest crops, grass, plucks fruits as well dry fruits, squeeze oil from the local nuts, also perform weaving, animal husbandry, poultry, local business, vegetable production, pulses, cattle breeding, thrashing, stone masonry and mines practices as well as labour to the landlord and contractor. The Polyandrous women of tribal areas have more responsibility to manage. There family along with all the work as well as to satisfy her all husband without erupting dispute or violence. Such customs on society provides an opportunity to study a polyandrous group more closely in affinity and harmony with nature. The society here is ancient one and its description is found in Vedas, Gitas and Purans. The people of the area still love their culture beside modernity and globalization. Till nineties outsiders were not allowed without prior permission of authority as well as institution of devi and devta. As a result the Kinnaura community preserved their culture and tradition as compared to others ones tribe in India. Now the area is approachable through various transportation means such as buses, taxis as well as helicopter. Lots of researches are taking place by the social scientists in general as well as the psychologist in particular. In this study attempt has been made to explore the wellbeing of Polyandrous people living in the Kinnaur district of Himachal Pradesh in India. Therefore, the objective of the present study is to study the psychological wellbeing among polyandrous and monogamous socially disadvantaged tribal people of Kinnaur. The broader objective of the study is as follows:

4.2 METHOD

4.2.1 Study Area

The Study was conducted in different areas of district Kinnaur which is on National Highway-22 situated in Himachal Pradesh (India). More specifically, the present study has been conducted in villages Sapni, Sangala, Monne, Aakpa and Kalpa. The villages fall in Nichar, Kalpa and Pooh Blocks of the district Kinnaur. These villages are approximately 260 kilometer from Shimla and 100 kilometer from Rampur Bushahar approachable through public and private buses. The appropriate study site (Map) has been shown in the below mentioned figure (Show next page).

4.2.2 Sample

In the present study a sample of N = 80 subjects has been selected from aforesaid villages of the district Kinnaur. The sample selected in the present study was above 20 year of age. The selected sample of N = 80 subjects were divided into two sub–groups based on their marital status i.e. from polyandry (N=40) and from Monogamy (N=40) subjects. These subjects were again divided into two equal comparable half

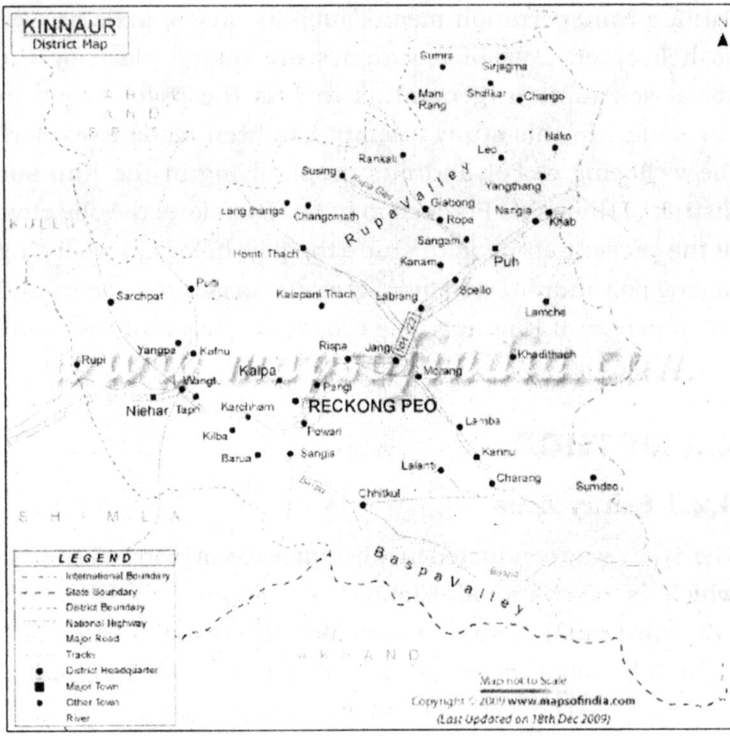

Fig 4.4.1 (a): Study area Description of Study Area

based on their gender status that comprises of N=20 in male groups as well as another N=20 in females groups. Finally, the subjects were again sub-divided into two comparable halves based on their caste that comprises of N=10 in under Scheduled Tribe category and remaining N=10 in Scheduled Caste categories. In all, there were eight groups with N=10 subjects in each that comprise of complete sample of 80 subjects of aforesaid characteristics. A 2 × 2 × 2 factorial design was used and the description is as follows:-

4.2.3 Design

A 2 x 2 x 2 Factorial Designs

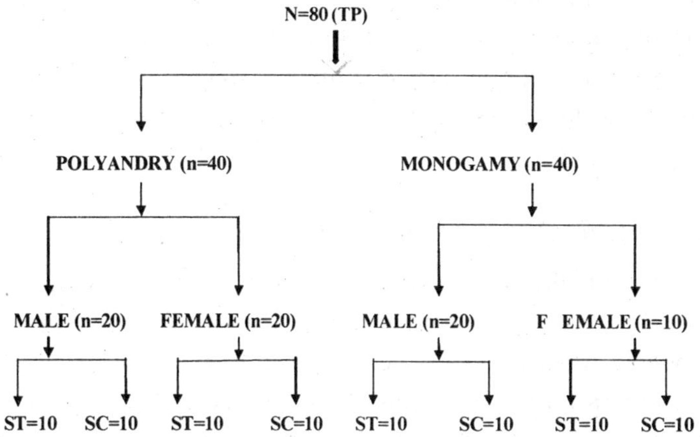

Notations-: TP= Tribal People; ST= Scheduled Tribe; SC= Scheduled Caste

4.2.4 Measure

In this study both qualitative and quantitative measures were used. The descriptions of both are as follows:

4.3 QUALITATIVE METHOD

4.3.1 Observation

Observation is a fundamental aspect of any science and has been playing an essential part in the development of psychology as a scientific discipline. Observation involves the noting and recording of events and human behavior without formal manipulation of variables operating in the events.

4.3.2 Interview

Interview is a formal or informal meeting between two people or among a group of people for the purpose of obtaining information about something in particular. It is a face to face conversation which is always goal directed. It is considered as a successful tool in social sciences research.

4.3.3 FCM: (Fuzzy Cognitive Mapping)

Individuals store their own perception of the reality through mental maps. In cognitive maps, the stored concepts are decoded, analyzed and clearly structured through cause and effect relationships, and this is significantly useful both for individual aims (through understanding of complex issues) and in group situations. A cognitive map is constituted by nodes, which represents the concepts and are connected to each other by links (also called edges). The edges are directed to show the directions of the cause-effect relationships. In the fuzzy cognitive maps, developed initially by Kosllo in 1986, cognitive mapping is a pure qualitative technique: the use of Fuzzy Logic introduces some quantification aspects. A successful application of the Fuzzy cognitive maps as knowledge mapping systems is the study of Ozesmi in 1999 (Ozesmi, 2006).

4.3.4 Photography

Photography is the process of forming stable or permanent visible images directly or indirectly by the action of light or other forms of radiation on sensitive surfaces.

4.4 QUANTITATIVE MEASURES

4.4.1 PGI- Health Questionnaire N-1

It was developed by Verma, Wig and Pershad in 1985 that has two sections. Section A consists of 16 questions and section B of 22 questions. This test is a short simple and moderate in difficulty levels already used by different researchers on illiterate and less educated people generally, and discriminatory power. Reliability of the test was examined using 'test-retest' and 'split half' methods signify 0.88 and 0.86 respectively. Subject were asked to put a tick mark against the question in the column with which they were agree. The number of ticks in section 'A' and 'B' indicated the respective scores, which were then to be added to give a total score. Higher the score, according to Verma, et al. (1985) indicates better will be chance of developing psychiatric symptoms. The score ranges from minimum of zero to a maximum of 16 in part A and 22 in part B as well as 38 in total. More the score, more neurotic the person will be.

4.4.2 Polyandry Attitude Inventory

It was developed by Sunder Kala Negi in 2010. This inventory comprises of 40 items divided into two equivalent parts. The Part-A of the inventory deals with the causes whereas the Part-B deals with the consequences of polyandrous marriage. Each section consists of 20-20 items with Yes and No options. The score ranges from minimum of zero to a maximum of 20 in each section. This inventory has been prepared in order to understand more deeply about the attitude of tribal people towards polyandry system. More the score more will be the attitude towards causes and consequences of the polyandry marriage.

4.4.3 State-Trait Anxiety Inventory (STAI-HX-1)

Based upon Spielberger's work on anxiety, Sharma (1988) has clarified that a temporal sequence was involved in the relationship between stress and anxiety, where stress denotes the stimulus properties of a situation and threat refers to an individual's perception of a situation as more or less dangerous or threatening to him. These temporal events arouse state anxiety among the population. The concept of state anxiety implies an intensity dimension with different qualitative characteristics of anxiety states at different levels of intensity. The tool of state anxiety used in the present study in the present study was the original work of Spielberger, Sharma and Singh (1973) that consists of 20 items in Hindi language. Each item measures a response on 4 point scale. Half of the items are scored conventionally whereas another half i.e. item number 1,2,5,8,10, 11, 15,16, 19, 20 are scored in reverse direction. The score range from minimum of 20 to a maximum of 80. It has been standardized on north India population and is simple to administer.

4.5 PROCEDURE

Kinnaur is situated in the Eastern district, second largest area-wise and 2nd least populated district of the Himachal Pradesh, approximately 260 Kms. away from Shimla. The gate to enter this district is visible at Chaura village. In this district there exist various hierarchies of groups with various castes, income, and ethno linguistic and socio-cultural groups. The objective of the present study is to study the psychological well-being among polyandrous and monogamous socially disadvantaged tribal people of Kinnaur. The study has been conducted on a sample of N =80 subjects. The sample selected in the present study was above 20 year of age. A sample of N = 80 subjects were divided into two sub–groups based on their marital status i.e. from polyandry (N=40) and Monogamy (N=40) subjects were selected. These subjects were again divided into two comparable half based on their Gender status that comprises of N=20 in male as well as another N=20 in females groups. Finally, the subjects were sub-divided into two comparable halves based on their caste that comprises of N=10 in Scheduled Tribe as well as another N=10 in Scheduled Caste groups. In all, there were eight groups with N=10 subjects in each that comprise of complete sample of 80 subjects of aforesaid characteristics. Thus A 2 × 2 × 2 factorial design was used. These eight groups of subjects were assessed with the help of qualitative and quantitative measures The qualitative measure include, Field observation, Interview, Fuzzy Cognitive Mapping and Photography whereas the quantitative measure include Polyandry Attitude Inventory, PGI Health Questionnaire, and State-anxiety inventory. The quantitative measure is self-report in nature that was be administered by the researcher herself by asking open ended question, from the participants. Since the researcher herself deals with Kinnaur district helped a lot, in asking questions in her own Language in carrying out numerical for this manuscript. Besidc this the researchers did her pilot study and asked various people about polyandry system in this district. Later on Gram Pradhan, educated youths as well as elderly literate and illiterate people were asked to know deeply about polyandrous attitude of the people in the concerned villages. The researcher entered in the particular village, established warmth atmospheres by addressing the objective of the present study. The ideas elicited by the subjects were welcomed by the researchers and they were encouraged to explain more about the system. The researcher used observation, interview, fuzzy cognitive mapping as well as photography to record the polyandrous system. Through qualitative analysis their view were record and described in the result section. Beside this the subjects were analyzed though PGI Health Questionnaire, and State-Trait Anxiety Inventory in order to record qualitative data about the well being too so as to verify the dependent variable justified and scientifically. Their scores were tabulated and means were compared and analyzed by performing three way analysis of variance. The description of the result has been presented in the next chapter.

TRIBAL POLYANDROUS AND MONOGAMOUS PEOPLE

In this chapter attempt has been made to evaluate the various dimension of well-being among polyandrous and monogamous, male and female belonging to upper and lower castes from district Kinnaur in Himachal Pradesh (India). In the below mention paragraphs an attempt has been made to record and tabulate the scores of these people on different areas of the well-being that include physical, social and psychological dimension. In the first instance the PGI-Health questionnaire area-A of the polyandrous and monogamous peoples have been compared and the description is as follows:-

6.1 PHYSICAL HEALTH AMONG TRIBAL POLYANDROUS AND MONOGAMOUS PEOPLE

Among the tribal communities, it is quite difficult to study the problems of health isolated from other aspect of their culture. In primitive societies the practice of medicine was associated with religion, magic and astrology. The different socio-cultural activities revolve around gods for their health and diseases. Tribal communities dwelling inaccessible and remote areas, always are confronted with want, hunger, poverty awfully uncongenial environment and natural calamities those are bond to the superstitions. The PGI-Health questionnaire was given to the tribal population and then score was tabulated as follows on the physical aspect of the neuroticism.

Table 6.6.1: Average scores of Scheduled Tribe and Scheduled Caste, Male-female, Polyandry and Monogamy People in Physical Aspect of Neuroticism

Polyandry

	M	F	AV
ST	5.5	8.1	6.8
SC	7.4	8.2	7.8
AV	6.45	8.15	7.3

Monogamy

	M	F	AV
ST	3.9	5.9	4.9
SC	8.3	9.5	8.9
AV	6.1	7.7	6.9

Notation: ST/SC = Scheduled Tribe/Cast; M/F = Male/Female; AV = Average scores

From the table 6.6.1 it is quite evident that the average score of polyandrous Scheduled tribe people on physical aspects of neuroticism (Area-A) was 6.8. And average score of Scheduled caste polyandrous people was 7.8. The difference between the two is (7.8-6 .8 = 1.0). Thus it shows slight difference between them. Polyandrous Scheduled caste people have been suffering little more with health problem than Polyandrous Scheduled tribe people in the tribal areas of Kinnaur.

The average score of polyandrous male on physical aspects of neuroticism was 6.45 whereas the polyandrous female as 8.15. The difference between the two sets of score is (8.15-6.45=2.70). It is quite clear that the polyandrous females have suffering from more heath problems. Similarly the average score of Scheduled Tribe polyandrous male was 5.5 as compared to scheduled caste polyandrous male that was found 7.4 and the difference between the two is (5.5 – 7.4 = 1.9). So it is quite clear that scheduled caste polyandrous male have more health problem than Scheduled Tribe polyandrous male. Further average score of polyandrous female scheduled tribe was 8.1 as compared to scheduled caste polyandrous female that was 8.2. The difference between both is (8.2-8.1) = 0.1. So this difference shows meager/negligible difference between them. Moreover, polyandrous females have been slightly poor in health behavior.

The average score of male polyandrous scheduled tribe as 5.5 and female polyandrous scheduled tribe as 8.2, so the difference between both is (8.2-5.5) = 2.7. So it is quite clear from the table that the male polyandrous scheduled tribe should less health problem than there counterpart female polyandrous scheduled tribe. The average score of scheduled caste polyandrous male as 7.4 whereas the average score of female scheduled caste polyandrous as 8.2. The difference is (8.2.-7.4) = 0.8 it shows that male scheduled caste polyandrous has less health problem than female scheduled caste polyandrous. Further, the average score of scheduled tribe polyandrous male as 5.5 and average score of female scheduled caste polyandrous as 8.2 now on comparison the difference came out as (8.2-5.5) = 2.7. Thus there is clear difference between them and it show that male scheduled tribe polyandrous has less health problem than the female scheduled caste polyandrous. Finally the average score of female scheduled tribe was 8.1 whereas the average score of male polyandrous scheduled caste was 7.4. The difference between the two is (8.1-7.42) = 0.7. The score has been showing slight difference between them. Thus scheduled caste male polyandrous has less health problem than scheduled tribe female counter part.

The average score of monogamous scheduled tribe on physical aspects of neuroticism as 4.9 whereas the average scores of scheduled caste monogamous people on physical aspects of neuroticism as 8.9. The difference between two is (8.9-4.9)=4.0. Thus the picture is so clear that the monogamous scheduled caste people have been suffering from more health problem than scheduled tribe monogamous. Similarly the average score of monogamous male on physical aspects of neuroticism as 6.1 and of monogamous female as 7.7. Now on comparison the difference came out as (7.7-6.1)=1.6. Thus there is slight difference between them and the monogamous female has been suffering with more physical health problem than that of monogamous male. Average score of scheduled tribe monogamous male as 3.9 as compared to scheduled caste male monogamous that as 8.3 and difference emerged out to be (8.3-3.9) = 4.4. So it is very much clear that male scheduled caste monogamous has envisaged more health problems than scheduled tribe male monogamous. Further the average score of female monogamous scheduled tribe as 5.9 and the average score of female monogamous scheduled caste as 9.5 and the difference between both as (9.5-5.9) = 3.6, showing scheduled caste monogamous female as suffering with more health problems than scheduled tribe monogamous females. Moreover, polyandrous females have been found slightly poor in health behavior.

In the same time the average score if male Scheduled tribe monogamous as 3.9 and the average score of female Scheduled tribe monogamous as 5.9 and the difference between both emerged out as (5.9-3.9) = 2.0. So it is quite evident from the treatment mean that the female scheduled tribe monogamous showed more health problems as compare to male Scheduled tribe monogamous. The average score of male monogamous scheduled caste as 8.3 and the average score of female monogamous scheduled caste as 9.5. The difference between the two is (9.5-8.3) = 1.2. Thus it shows that the monogamous female scheduled caste has been suffering more with physical health problems than male scheduled caste monogamous.

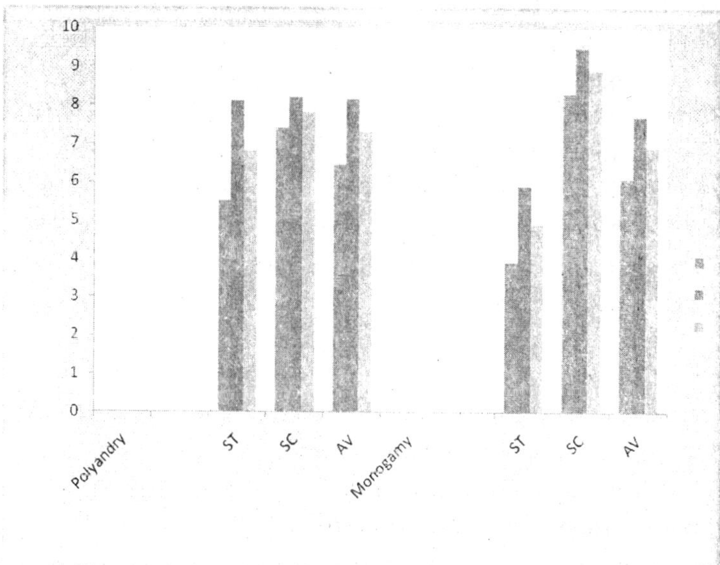

Figure 6.6.1.a: The Average scores of scheduled tribe and scheduled caste, male-female, polyandrous and monogamous population on physical Aspect of neuroticism

Notation: ST/SC = Scheduled Tribe/Cast; M/F = Male/Female; AV = Average scores

Similarly the average score of male scheduled tribe monogamous as 3.9, and the scheduled caste monogamous female as 9.5 the difference between both emerged out to be (9.5-3.9) = 5.6, which gives us clear picture about the scheduled caste monogamous females on ill indication. The average score of male scheduled caste monogamous was 8.3, the average score of female Scheduled tribe 5.9 and the difference between them is (8.3-5.9) = 2.4 shows that scheduled caste male have more health problems, than their counterpart.

Further, the average score of polyandrous Scheduled tribe male-female as 6.8, the average score of polyandrous scheduled caste male and female as 7.8. The difference between both is (7.8-6.8) = 1.0 reflect the light about the polyandrous scheduled tribe Male and female about less health problem than their counterparts. In the same fashion the average score of polyandrous scheduled tribe as 6.8 whereas average score of monogamous scheduled tribe as 4.9 and the difference is (6.8-4.0) = 1.9. It pinpoints that the monogamous scheduled tribe has less health problem, than their counterparts. The average score of polyandrous scheduled tribe as 6.8, average score of monogamous scheduled caste as 8.9 and the difference of polyandrous scheduled caste as 7.8, average score of monogamous scheduled tribe as 4.9 and the difference is (7.8-4.9) = 2.9. It shows that monogamous scheduled tribes have less health problems than polyandrous scheduled caste. The same trend was observed in scheduled caste polyandrous people whose average score as 7.8 than their counterparts scheduled caste monogamous people (8.9) and the difference came out as (8.9-7.8) = 1.1. Than finding shows that the monogamous Scheduled tribe has witnessed more health problem than polyandrous scheduled caste. Finally the average score of monogamous Scheduled tribe as 4.9, average score of monogamous scheduled caste as 8.9 and the difference sought out to be (8.9-4.9)=4.0 evidently showing that the monogamous scheduled caste has more physical health problems than monogamous scheduled tribe, counterparts.

6.2 MENTAL HEALTH AMONG TRIBAL POLYANDROUS AND MONOGAMOUS PEOPLE

To assess the psychological health among the tribal polyandrous and monogamous people across India and Himachal Pradesh have been proving as the innovative and pioneer area for researches in these areas among the sociologists as well as psychologists. Even after several programmes organized by NGOs and Government organization in the tribal areas by way of educational camps/ through media, audio-programmes, the psychological health among the tribal in Himachal Pradesh is very poor being the very reason of illiteracy and strong belief in local Gods (Devi-Devtas of the area). However, there is gradually and steadily growth and progress in the tribal areas.

Table 6.6.2: The Average scores of Scheduled Tribe and Scheduled Caste, Male-female, Polyandrous and monogamous population on Psychological Aspect of Neuroticism

Polyandry

	M	F	AV
ST	6.6	10.7	8.65
SC	13.1	11.9	12.5
AV	9.85	11.3	10.57

Monogamy

	M	F	AV
ST	5.0	9.0	7.00
SC	11.9	15.0	13.00
AV	8.45	12.0	10.22

Notation: ST/SC = Scheduled Tribe/Cast; M/F = Male/Female; AV = Average scores

From the table 6.6.2 it is quite clear that the average score of polyandrous Scheduled tribe on psychological aspects of neuroticism (Area-B) as 8.65 whereas the average score of polyandrous scheduled caste people in the same emerged out to be the 12.5. and the difference between the two as (12.5-8.65) = 3.85. Thus the score show slight difference between two sets of scores. The finding revealed that the polyandrous scheduled caste people have been suffering with more psychological problem, than their counterpart. Similarly, the average score of polyandrous male on psychological aspects of neuroticism as 9.85 and of polyandrous female as 11.3. Now on comparison the difference came out as (11.3-9.85) = 1.45. Thus, there was slight difference between two sets of score which shows that the female polyandrous has been suffering more psychological problems than the polyandrous males.

Further, the average score of Scheduled tribe polyandrous male was 6.6, female polyandrous Scheduled tribe as 10.7 and difference between both the two as (10.7-6.6) = 4.1. Thus Scheduled tribe female polyandrous are suffering with more psychological problem, than comparison group. The average score polyandrous scheduled caste female of as 13.1 and the average score of polyandrous scheduled caste female as 11.9; the difference between the two is (13.1-11.9) = 1.1. Thus it was showing that polyandrous scheduled caste male has been suffering from more psychological problem as compared to the polyandry scheduled caste female. Similarly the average score of polyandrous male scheduled caste as 13.1 and female as 6.6. The difference between the two came out to be (13.1-6.6) = 6.5. Thus it shows very much difference which helps in concluding that the male scheduled caste has been suffering with more health problem than male Scheduled tribe polyandrous counterpart.

Similarly, the average score of polyandrous Scheduled tribe female as 10.7 and the average score of female polyandrous Scheduled tribe as 11.9, and the difference between the both is (11.9-10.7) = 1.2. Thus there is slight difference between them and thus scheduled caste polyandrous was female suffering more with psychological health problem than their counterpart. Further, the average score of Scheduled tribe male polyandrous as 6.6 and the average score of scheduled caste polyandrous female as 11.9. The difference between both is (11.9-6.6) = 5.3 and it shows that scheduled caste female polyandrous population is suffering more with psychological problems. The average score of Scheduled tribe male polyandrous is 6.6 and the average score of scheduled caste polyandrous female is 11.9, the difference between both is (11.9-6.6) = 5.3. Thus it shows clear difference between them and the scheduled caste female polyandrous population has been suffering more psychological problem than Scheduled tribe male polyandrous counterparts. Finally the average scores of female Scheduled tribe is 10.7 whereas the average scores of male polyandrous scheduled caste is 13.1. The difference between the two is (13.1-10.70) = 2.3. Thus it is shows that there emerged out to be a quite noticeable difference between them and the male scheduled caste polyandrous showed more psychological problems than female Scheduled tribe polyandrous, counterpart.

The average score of monogamous scheduled tribe on psychological aspects of neuroticism (Area-B) as 7.00 whereas the average score of scheduled caste monogamous people on the same as 13.00 and the difference between two sets of score came out to be (13.00-7.00)= 6.0. Thus the picture was so clear that the monogamous scheduled caste people have suffered a lot from health behaviour problems than

Scheduled tribe monogamous population.

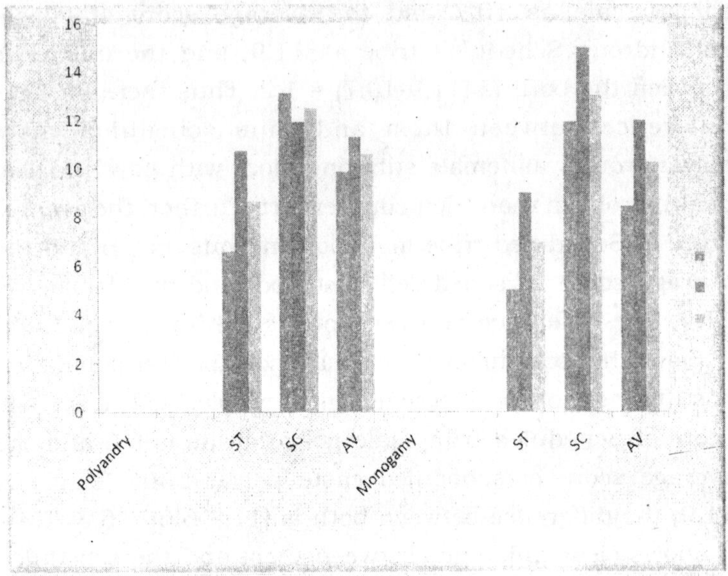

Figure 6.6.2.b: The Average scores of scheduled tribe and scheduled caste, male-female, polyandrous and monogamous population on psychological aspect of neuroticism

Notation: ST/SC = Scheduled Tribe/Cast; M/F = Male/Female; AV = Average scores

The average score of monogamous male on psychological aspects of neuroticism as 8.45 and the monogamous female as 12.0. Now on comparison the difference came out as (12.0-8.45) = 3.55. Thus there is clear difference between them and showed that monogamous female is suffering with more psychological problem as compared to their counterpart monogamous male. Similarly the average score of Scheduled tribe monogamous male as 5.0 and scheduled caste male monogamous as 11.9 and the difference (11.9-5.0) = 6.9. Thus the picture is showing that the male scheduled tribe monogamous has comparatively less psychological problem than male scheduled caste monogamous counterpart. The average score of female monogamous scheduled tribe as 9.0 and the female monogamous scheduled caste as 15.0, the difference between both was (15.0-9.0) = 6.0. So it shows very clear picture that scheduled caste monogamous female has more psychological problem than scheduled tribe female monogamous, counterpart.

The average score of male scheduled tribe monogamous as 5.0, the average score of female scheduled tribe monogamous as 9.0 and the difference between both is 9.0-5.0= 4 showing that female scheduled tribe monogamous has reported more health problems than male scheduled tribe monogamous. Similarly, the average score of male monogamous scheduled caste as 11.9 the female monogamous scheduled caste as 15.0, and the difference between two was (15.0-11.9) = 3.1. Thus it shows that the monogamous female scheduled caste is suffering with more psychological problem than scheduled caste male monogamous population. The average score of male scheduled tribe monogamous as 5.0 and the average score of scheduled caste female monogamous as 15.0 and the difference between both came out on (15-5) = 10. Thus this great difference shows that scheduled caste female monogamous has very high health problems than scheduled tribe male monogamous counterpart. The average score of male scheduled caste monogamous people as 11.9, than the average score of female scheduled tribe monogamous as 9.0 and the difference between them came out (11.9-9.0) = 2.9. So it is quite clear that male scheduled caste has more psychological problem than female scheduled tribe monogamous.

Further, the average score of polyandrous Scheduled tribe male female as 8.65, the average score of polyandrous scheduled caste male female as 12.5 and the difference between both is (12.5-8.650) = 3.85. So the difference between them shows that scheduled caste polyandry (male & female) has more psychological problem than Scheduled tribe polyandrous population. The average score of polyandrous scheduled tribe as 9.65, average score of monogamous scheduled tribe as 7.00 and the difference came out (8.65-7.00) = 1.65. It shows that the monogamous Scheduled tribe has less heath problems than their counterpart group. Similarly the average score of polyandrous Scheduled tribe as 8.65, average score of monogamous scheduled caste as 13.00 and the difference between both is (13.00-8.65) = 4.30. It shows that the polyandrous Scheduled tribe population has less health problem than comparison group. The average score of polyandrous scheduled caste as 12.5, average score of monogamous scheduled tribe as 7.00 and the difference is (12.5-7.00)= 5.5. It shows monogamous scheduled tribe has less health problem than the polyandrous scheduled caste. The average score of polyandrous scheduled caste as 12.5, average score of monogamous scheduled caste as 13.00 and the difference between them as (13.00-12.5) = 5.00. It shows that the monogamous scheduled tribe has repeated scant health problem than polyandrous scheduled caste population. Finally the average score of monogamous scheduled tribe as 7.00 and monogamous scheduled caste as 13.00 and the difference is (13.00-7.00) = 6. It shows that the monogamous scheduled caste has been suffering with more psychological problems than monogamous scheduled tribe counterpart.

6.3 GENERAL HEALTH AND WELL-BEING AMONG TRIBAL POLYANDROUS AND MONOGAMOUS PEOPLE

As far as health is concerned, it is a very broad term to define. It is a state of positive well being and feeling of internal happiness is an individual. It is more than the state and prevails in person free of injuries.In Indian traditional culture, health is viewed as individual embedded in environment. The traditional health systems of Indian understood health and well-being as interaction of an individual's mental, physical, social and spiritual life. The tribal in the State of Himachal Pradesh have poor health; mentally as well as physically due to the climatic conditions, hardworking nature and ignorance of their health problems. The topography, atmospheric pressure, strong sun-light, less vegetation, low rains and heavy snow fall are the main reasons behind this, with the ignorance of care and medicines.

Table 6.6.3: Average scores of scheduled tribe and scheduled caste, male-female, polyandrous and monogamous population on physical and psychological aspect of neuroticism

Polyandry

	M	F	AV
ST	6.05	9.4	7.72
SC	10.25	10.05	10.15
AV	8.15	9.73	8.94

Monogamy

	M	F	AV
ST	4.45	7.45	5.95
SC	10.1	12.25	11.17
AV	7.27	9.85	8.56

Notation: ST/SC = Scheduled Tribe/Cast; M/F = Male/Female; AV = Average scores

From the table 6.3.1 it is quite clear that the average score of polyandrous scheduled tribe on physical and psychological aspects neuroticism as 7.72 whereas the average score of scheduled caste polyandrous people on physical and psychological aspects of neuroticism as 10.15. The difference between the two is (10.15-7.72) = 2.43. Thus it was shows that the polyandrous scheduled caste people has been suffering through more psychological and physical problems than polyandrous scheduled tribe people in the tribal areas of the Kinnaur.

The average score of polyandrous male on physical and psychological aspects of neuroticism as 8.15 whereas the polyandrous female as 9.73. Now on comparison the difference came as (9.73-8.15) = 1.58. Thus there is slight difference between them it shows that females are suffering with more depression and physical problem than the polyandrous males. Similarly the average score of scheduled tribe polyandrous male as 6.05 as compared to scheduled caste polyandrous male that as 10.25 and the difference between two is (10.25-6.05) = 4.20. So it is clear that scheduled caste polyandrous male has more health problems than their counterpart scheduled tribe polyandrous male. Further the average scores of scheduled tribe polyandrous female as 9.4 as compared to scheduled caste polyandrous female that as 10.05. The difference between both is (10.25-9.4) = 0.65. Thus there is negligent difference between them and polyandrous scheduled caste female has more physical and psychological health problem.

Similarly the average score of male polyandrous Scheduled tribe as 6.05 and female scheduled tribe as 9.4 so difference between both is (9.4-6.05) =.3.35. So it is quite clear from the table that the male scheduled tribe polyandrous has less health problem than their counterpart female scheduled tribe polyandrous. The average score of scheduled caste male polyandrous as 10.25 and the average of scheduled caste polyandrous female as 10.05. The difference is (10.25-10.05) = .20 so there is slight difference between them. The average score of scheduled tribe polyandrous male as 6.05 and average score of female scheduled caste polyandrous as 10.05. Now on comparison the difference came as (10.05-6.05) = 4.0. The score has been shows clear difference between them. Thus scheduled tribe polyandrous male has low physical and psychological health problem than scheduled caste polyandrous female. Finally the average scores of scheduled tribe polyandrous female as 9.4 whereas the average scores of scheduled caste polyandrous male as 10.25. The difference between the two is (10.25-9.4) = 0.85. Thus it shows that there is slight difference between them and scheduled caste male polyandrous has more health problem than their counterpart female polyandrous scheduled tribe.

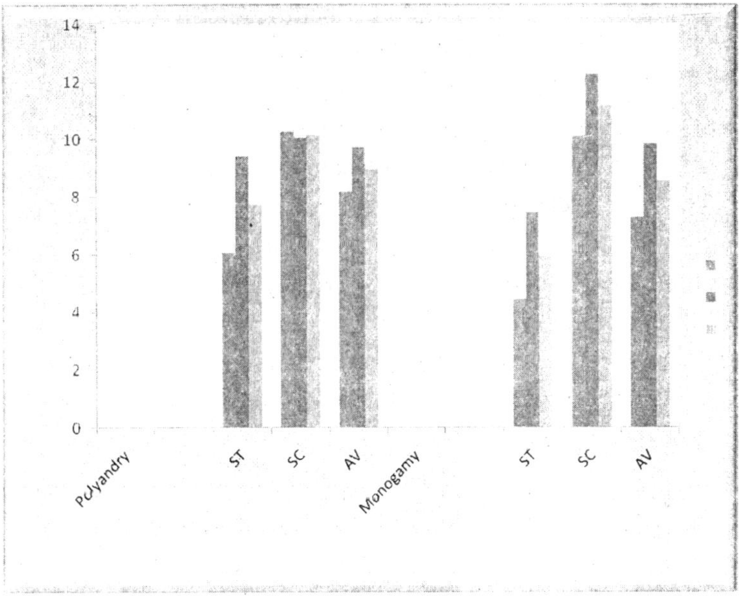

Figure 6.6.3.c: The average scores of scheduled tribe and scheduled caste, male-female, polyandrous and monogamous population on physical and psychological aspect of neuroticism

Notation: ST/SC = Scheduled Tribe/Cast; M/F = Male/Female; AV = Average scores.

The averages score of monogamous scheduled tribe on physical and psychological aspects of neuroticism as 5.95 whereas the average scores of scheduled caste monogamous people on psychological and physical aspects of neuroticism as 11.17. The difference between two is (11.17-5.95) = 5.22. Thus there picture is so clear that the monogamous scheduled caste people has suffering from more health problem than scheduled tribe monogamous. Finally the average score of monogamous male on physical and psychological aspects of neuroticism as 7.27 and monogamous female as 9.85. Now on comparison the difference came as (9.85-7.27) = 2.58. Thus monogamous female are suffering with more physical and psychological problems. Average score of scheduled tribe monogamous male was 4.45 as compared to scheduled caste monogamous male were 10.1 and the difference sought out to be (10.1-4.45) = 5.65. So it is very much clear that scheduled caste monogamous male has more physical and psychological problem than their counterpart scheduled tribe monogamous male. The average score of scheduled tribe monogamous female as 7.45 and the average score of scheduled caste monogamous female as 12.25. The difference between both emerged out to be (12.25-7.45) = 4.80, which gives us clear picture about the scheduled caste monogamous female ill health indication than Scheduled tribe monogamous females.

Similarly the average score of Scheduled tribe monogamous male as 4.45, the average score of scheduled tribe monogamous female as 7.45 and the difference between both is (7.45-4.45) = 3.0. So it is quite evident from the treatment mean that the female scheduled tribe monogamous showed more health problem as compare to male scheduled tribe monogamous. The average score of scheduled caste monogamous male as 10.1 and the average score of scheduled caste monogamous female as 12.25. The difference between the two is (12.25-10.1) = 2.15. Thus it shows that the monogamous female scheduled caste was suffering through more physical and psychological problem than their counterpart male scheduled caste monogamous. Finally, the average score of male scheduled tribe monogamous as 5.0 and the average score of scheduled caste monogamous female as 15.0. The difference between both is (15.0-5.0) = 10. Thus the picture is so clear that female scheduled caste monogamous has high physical and psychological problems. The average score of male scheduled caste monogamous as 10.1 and female scheduled tribe monogamous as 7.45 and the difference between both is (10.1-7.45) = 2.65. It clearly shown that monogamous scheduled caste male has more physical and psychological problems than scheduled tribe female monogamous.

The average score of polyandrous scheduled tribe (male and female) as 7.72. The average score of polyandrous scheduled tribe (Male and female) as 10.15 and the difference between both is (10.15-7.72) = 2.45. So it is quite clear that polyandrous scheduled tribe has low health problem. The same trend was observed in scheduled tribe polyandrous people whose average score was 7.72 than their counterpart scheduled tribe monogamous people 5.95 and the difference came out as (5.95-7.92)=1.77. The finding shows that the polyandrous scheduled tribe has witnessed more physical, psychological health problem than monogamous scheduled tribe. The average score of scheduled tribe polyandrous as 7.72, average score of scheduled caste monogamous as 11.17 and the difference between sought out to be (11.17-7.72) = 3.45. Evidently showing that the monogamous scheduled tribe has less health problems than scheduled caste monogamous their counterparts. Finally the average score of scheduled caste polyandrous as 10.15, average score of scheduled tribe monogamous as 5.95 and difference is (10.15-5.95) = 4.20. So it is clear that monogamous scheduled tribe has less health, physical, psychological problem than polyandry scheduled caste people.

The average score of scheduled caste polyandrous as 10.15. Average score of scheduled caste monogamous as 11.17 and the difference is (11.17-10.15) = 1.02. So it is quite clear that polyandry scheduled caste has low health, psychological and physical problem than monogamous scheduled caste people. Finally the average score of monogamous scheduled caste as 11.17, the difference between both is (11.17-5.95) = 5.22. So it is very highly clear that monogamous scheduled caste has more physical, psychological problem than scheduled tribe monogamous people.

6.4 ATTITUDE OF POLYANDROUS AND MONOGAMOUS TRIBAL PEOPLE TOWARDS THE CAUSES OF POLYANDRY

The tribal people have almost beliefs and faith in devtas comparatively their beliefs, attitude, attention and care towards themselves are not so positive. The people of these tribal's' areas are very hard working and there life style is very much difficult and most the time they ignore daily care and hygiene. although Gupta (1980) feels that fate of the individuals and community at large depends on their relationships with unseen forces, which intervenes in human affairs, yet it need to modify If men offended them, the mystical's poor punishes them by sickness, death or natural calamities so to a tribal mind, the real enemies of human health and prosperity are the Gods and the dead. The quality of life of tribal area is not so good because of – pressure of physical work, alcohols, malnutrition and the climatic conditions.

Table 6.6.4: Average scores of Scheduled Tribe and Scheduled Caste, Male-female, Polyandrous and Monogamous People on Causes of Polyandry.

Polyandry

	M	F	AV
ST	13.7	15.2	14.45
SC	12.7	14.9	13.8
AV	13.2	15.05	14.125

Monogamy

	M	F	AV
ST	11.4	12.0	11.7
SC	13.7	14.8	14.25
AV	12.55	13.4	12.975

Notation: ST/SC = Scheduled Tribe/Cast; M/F = Male/Female; AV = Average scores

The average score of polyandrous scheduled tribe on the causes of polyandry as 14.45 and average score of scheduled caste polyandrous as 13.8 from average score. The difference between the two is (14.5-13.8) = .7. Thus it shows slight difference between them and polyandrous scheduled caste people low causes of polyandry than polyandrous

scheduled tribe. The average score of polyandrous male between (scheduled caste, scheduled tribe) on the same was 13.2 and whereas the polyandrous female both (scheduled caste, scheduled tribe) has 15.05 the difference between two sets of score is (15.05-13.02) = 1.85. Thus it is clear that the female polyandrous has low causes than polyandrous male. The average score of scheduled tribe polyandrous male as 13.7 as compare to scheduled caste male polyandrous that as 12.7 the difference between two is (13.7-12.7)= 1.0. So it quite clear that scheduled caste male has high causes than Scheduled tribe male polyandrous. The average score of polyandrous female scheduled tribe was 15.2 as compare to scheduled caste female polyandrous that was 14.9. The difference between both is (15.2-14.9) = .3. So this difference shows major negligible difference between them.

The average score of male polyandrous scheduled tribe as 13.7 and female scheduled tribe polyandrous as 15.2 so the difference between both is (15.2-13.7)= 1.5. So it quite clears that male polyandrous scheduled tribe has more causes of polyandry than high quality of life than female polyandrous scheduled tribe. The average score of scheduled caste male polyandrous as 12.7 and the average of female scheduled caste polyandrous as 14.9. The difference is (14.9-12.7) = 2.2 and it shows that male scheduled caste polyandrous has more cause life than female scheduled caste polyandrous. The average score of scheduled tribe polyandrous male as 13.7 and average score of female scheduled caste polyandrous as 14.9. Now on comparison the difference came as (14.9-13.7) = 1.2. Thus there is clear difference between them and it shows that male Scheduled tribe polyandrous have more causes than female scheduled caste polyandrous. Finally the average score of female scheduled tribe as 15.2 where as the average score of male polyandrous as 12.7. The difference between two is (15.2-12.7) = 2.5. Thus the same is shown that scheduled caste male polyandrous have more causes than scheduled tribe female.

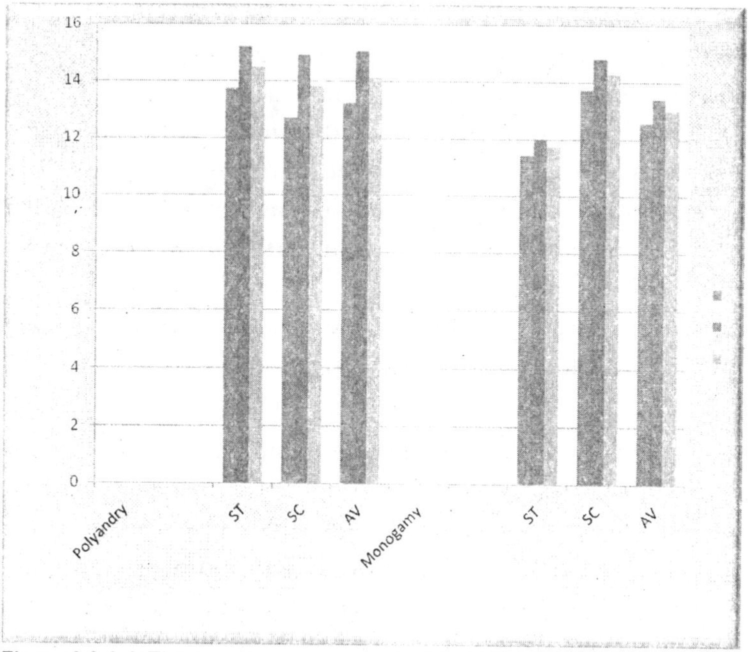

Figure 6.6.4.d: The average scores of scheduled tribe and scheduled caste, male-female, polyandrous and monogamous population on polyandry attitude inventory- (Area-A)

Notation: ST/SC = Scheduled Tribe/Cast; M/F = Male/Female; AV = Average scores

The Average score of monogamous scheduled tribe on causes of polyandry attitude Inventory was 11.7 where as the average score of scheduled caste monogamous people is 14.25 the difference between two is (14.25-11.7)=2.55. Thus the picture is clear

that the monogamous scheduled tribe has more causes than scheduled caste monogamous. The average score of monogamous male as 12.55 and of monogamous female were 13.4. Now on comparison the difference came as (13.4-12.55) = .85. Thus there is slight difference between them and monogamous male have more causes than female monogamous. The average score of scheduled tribe monogamous male was 11.4 as compare to scheduled caste male monogamous as 13.7 and the difference as (13.7-11.4) = 2.3. So it is very much clear that male scheduled caste less causes than male scheduled tribe monogamous. The average score of female scheduled tribe as 12.0 and the average score of female monogamous scheduled caste as 14.8, the difference between both is (14.8-12.0) = 2.8. Showing scheduled caste monogamous female was low causes than female scheduled tribe.

The average score of male scheduled tribe monogamous as 11.4. The average score of female Scheduled tribe monogamous as 12.0 and the difference between both is (11.4-12.00) = .6. So there is slight difference between them male and female Scheduled tribe has more causes than female scheduled tribe. The average score of male monogamous scheduled caste as 13.7 and the average score of female monogamous scheduled caste as 14.8. The difference between the two is (14.8-13.7) = 1.1. Thus it slows that the monogamous female scheduled caste has low causes of polyandry marriage. The average score of male scheduled tribe monogamous as 11.4, and the average score of scheduled caste female monogamous as 14.8 and the difference between both is (14.8-11.4) = 3.4. Thus the picture is so clear that female scheduled tribe monogamous has low causes than male scheduled tribe monogamous. The average score of male scheduled caste monogamous as 13.7, the average score of female scheduled tribe as 12.0 and the difference between them is (13.7-12.0) = 1.7, it shows that male scheduled caste have low causes than female scheduled tribe.

The average score of polyandrous scheduled tribe as 14.45 whereas average score of monogamous scheduled tribe as 11.7 and the difference is (14.45-11.70) = 2.75, it shows scheduled tribe monogamous has more causes than polyandrous scheduled tribe.

The average score of polyandrous scheduled tribe as 14.45. The average score of monogamous scheduled caste as 14.25 and the difference is (14.45-14.25) = .20. It shows monogamous scheduled caste has high causes. The average score of polyandrous scheduled caste as 13.8, the average score of monogamous scheduled tribe as 11.7 and the difference is (13.8-11.7) = 2.1. Thus showing monogamous scheduled tribe has more causes of polyandry marriage. The average score of polyandrous scheduled caste as 13.8 and average score of monogamous scheduled caste as 14.25. Difference between both is (14.25-13.8)=.45. Thus showing monogamous scheduled caste has low causes than polyandrous scheduled caste. The average score of polyandrous (male, female, scheduled caste, Scheduled tribe) as 14.125, the average score of monogamous (male, females, scheduled caste, Scheduled tribe) as 12.975 and the difference is (14.125-12.975) = 1.150. It shows that monogamous has more causes of polyandrous life.

6.5 ATTITUDE OF POLYANDROUS AND MONOGAMOUS TRIBAL PEOPLE TOWARDS CONSEQUENCES OF POLYANDRY

The main focus of this study was to compare Polyandrous and monogamous people towards causes and consequences of polyandrous marriage. In polyandry marriages there are fraternal and non-fraternal polyandry in the society. In polyandry marriages are going on the females are being exploited and there health problems are increasing significantly. Due to globalization and rise in life style, the conditions of polyandry females are more critical which is moving towards poverty, unemployment, increase in population and illiteracy etc, the simplicity of the tribal females are also be mis-utilized and there hospitality are taken wrongly by the globalized modern people by attracting towards modernity and luxurious things.

Table 6.6.5: Average scores of Scheduled Tribe and Scheduled Caste, Male-female, Polyandrous and Monogamous People towards Consequences of Marriage.

Polyandry

	M	F	AV
ST	14.9	15.2	15.05
SC	13.0	13.1	13.05
AV	13.95	14.15	14.05

Monogamy

	M	F	AV
ST	11.0	11.3	11.15
SC	12.5	12.9	12.7
AV	11.75	12.1	11.925

Notation: ST/SC = Scheduled Tribe/Caste; M/F = Male/Female; AV = Average scores

The average score of polyandrous scheduled tribe on the polyandry attitude inventory area-B as 15.05 and average score of scheduled caste polyandrous as 13.05 from average score. The difference between the two is (15.05-13.05) = 2.0, thus it shows slight difference between them and polyandrous scheduled tribe people have low consequences than polyandrous scheduled caste.

The average score of polyandrous male between (scheduled caste, scheduled tribe) on the polyandry attitude inventory was 13.95 and whereas the polyandrous female both (scheduled caste, scheduled tribe) has 14.15 the difference between two sets of score was (14.15-13.95) = .2, thus it is clear that the female polyandrous has less consequences than polyandrous male. The average score of scheduled tribe polyandrous male was 14.9 as compare to scheduled caste male polyandrous that was 13.0 the difference between two is (14.9-13.0) = 1.9, so it quite clear that scheduled caste male has more consequences than scheduled tribe male polyandrous. The average score of polyandrous female scheduled tribe was 15.2 as compare to scheduled caste female polyandrous that was 13.1. The difference between both is (15.2-13.1) = 2.1, so this difference shows clear difference between them and scheduled caste polyandrous has more consequences towards polyandrous.

The average score of male polyandrous scheduled tribe as 14.9 and female scheduled tribe polyandrous as 15.2 so the difference between both is (15.2 14.9) = .3, so there is slight difference between both that male polyandrous scheduled tribe has more consequences than female polyandrous scheduled tribe. The average score of scheduled caste male polyandrous as 13.0 and the average of female scheduled caste polyandrous as 13.1. The difference is (13.1-13.0) = .1, it shows that there is negligent difference between both and male scheduled caste polyandrous has more consequences than female scheduled caste polyandrous. The average score of scheduled tribe polyandrous male as 14.9 and average score of female scheduled caste polyandrous as 13.1 now on comparison the difference came as (14.9-13.1) = 1.8. Thus there is clear difference between them and it shows that female scheduled caste polyandrous has more consequences than male scheduled tribe polyandrous .Finally the average score of female scheduled tribe as 15.2 where as the average score of male scheduled caste polyandrous as 13.0. The difference between two is (15.2-13.0) = 2.2. Thus the same is shown that scheduled caste male polyandrous has more consequences toward polyandry life than scheduled tribe female.

The Average score of monogamous scheduled tribe on polyandry attitude Inventory was 11.15 where as the average score of scheduled caste monogamous people is 12.7 the difference between two is (12.7-11.15)=1.55, thus the picture is clear that the monogamous scheduled tribe has more consequences than scheduled caste monogamous. The average score of monogamous male as 11.75 and of monogamous female were 12.1. Now on comparison the difference came as (12.1-11.750 = .35, thus there is slight difference between them and monogamous male has more consequences than female monogamous. The average score of scheduled tribe monogamous male was 11.0 as compare to scheduled caste male monogamous that was 12.5 and the difference as (12.5-11.0) = 1.5. So it is clear that male scheduled caste has less consequences than male scheduled tribe monogamous. The average score of female scheduled tribe as 11.3 and the average score of female monogamous scheduled caste as 12.9, the difference between both is (12.9-11.3) = 1.6. Showing scheduled caste monogamous female as low consequences of polyandry than female scheduled tribe.

The average score of male scheduled tribe monogamous as 11.0, the average score of female scheduled tribe monogamous as 11.3 and the difference between both is (11.3-11.0) = .3. So there is slight difference between them male scheduled tribe has more consequences than female scheduled tribe. The average score of male monogamous scheduled caste as 12.5 and the average score of female monogamous scheduled caste as 12.9. The difference between the two is (12.9-12.5) = .4, thus it is slowly that the monogamous female scheduled caste has low consequences of polyandry.

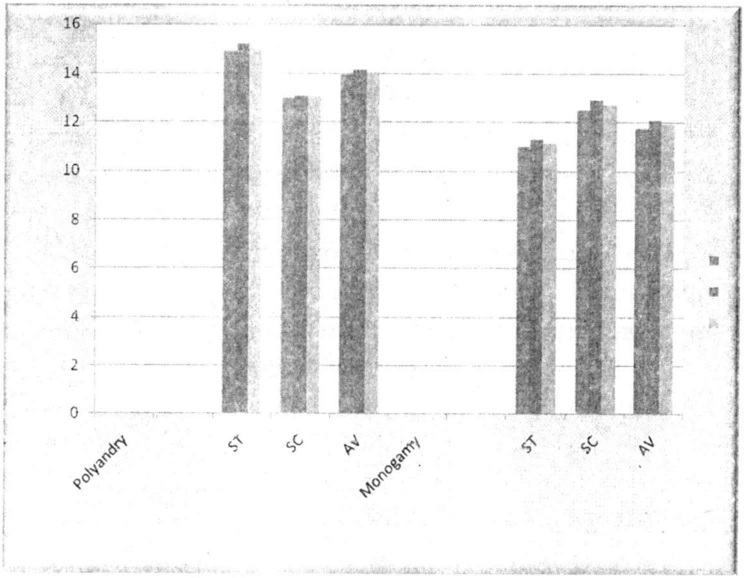

Figure 6.6.5.e: The Average scores of scheduled tribo and scheduled caste, male-female, polyandrous and monogamous population on polyandry attitude inventory-Area-B

Notation: ST/SC = Scheduled Tribe/Caste; M/F = Male/Female; AV = Average scores

The average score of male scheduled tribe monogamous was 11.0, the average score of scheduled caste female monogamous as 12.9 and the difference between both is (12.9-11.0) = 1.9. Thus the picture is so clear that female scheduled caste monogamous has low attitude towards consequences than male scheduled tribe monogamous. The average score of male scheduled caste monogamous as 12.5, the average score of female scheduled tribe as 11.3

and the difference between them is (12.5-11.3) = 1.2. thus it Shows that male scheduled caste has low attitude towards consequences than female scheduled tribe.

The average score of polyandrous scheduled tribe as 15.05, whereas average score of monogamous scheduled tribe as 11.15 and the difference is (15.05-11.15) = 3.90, it showing scheduled tribe monogamous has more consequences of polyandry than polyandrous scheduled tribe. The average score of polyandrous scheduled tribe as 15.05. The average score of monogamous scheduled caste as 12.7 and the difference is (15.05-12.7) = 2.35. It shows monogamous scheduled caste has more consequences. The average score of polyandrous scheduled caste as 13.05, the average score of monogamous scheduled tribe as 11.15 and the difference is (13.05-11.15) = 1.9. Thus showing monogamous scheduled tribe has high consequences. The average score of polyandrous scheduled caste as 13.05 and average score of monogamous scheduled caste as 12.7. Difference between both is (13.05-12.7) = .35. Thus showing monogamous scheduled caste has high consequences than polyandrous scheduled caste. The average score of polyandrous (male, females, scheduled caste , scheduled tribe) as 14.05, the average score of monogamous (male, female, scheduled caste, scheduled tribe) as 11.925 and the difference is (14.05-11.925) = 2.025. It shows that monogamous experience more consequences of polyandry marriage.

6.6 ATTITUDE OF POLYANDRY AND MONOGAMOUS TRIBAL PEOPLE TOWARDS CAUSES AND CONSEQUENCES OF POLYANDRY

Two most important factors which have caused the adoption of this practice by the people are scarcity of women and a desire to transmit and estate undivided (Y.S Parmar, 1975) but these polyandourous people faces some problems to compete with modern society due to globalization and rising in life style. The conditions of polyandry females are more critical which is moving towards poverty, unemployment, increase in population and illiteracy etc. The people of these tribal's' areas are very hard worker and their life style is very much difficult and most the time they ignore daily care and hygiene. The quality of life of tribal area is not so good because of high –pressure of physical work, alcohols, malnutrition and the climatic conditions.

Table 6.6.6: Average scores of Scheduled Tribe and Scheduled Caste, Male-female, Polyandrous and Monogamous Population on Polyandry attitude inventory Area–(A+B)

Polyandry

	M	F	AV
ST	14.3	15.2	14.75
SC	12.85	14.0	13.425
AV	13.575	14.6	14.087

Monogamy

	M	F	AV
ST	11.2	11.65	11.425
SC	13.1	13.85	13.475
AV	12.15	12.75	12.45

Notation: ST/SC = Scheduled Tribe/Caste; M/F = Male/Female; AV = Average scores

The average score of polyandrous scheduled tribe on the Polyandry Attitude Inventory both area was 14.75 and average score of scheduled caste polyandrous as 13.425 from average score. The difference between the two is (14.75-13.425) = 1.325. Thus it shows slight difference between them and polyandrous scheduled caste people have more causes and consequences than polyandrous scheduled tribe.

The average score of polyandrous male between (scheduled caste, scheduled tribe) on the this inventory was is 13.575 and whereas the polyandrous female both (scheduled caste, scheduled tribe) has 14.6 the difference between two sets of score is (14.6-13.575) = 1.025, thus it is clear that the female polyandrous has low causes and consequences of polyandry. The average score of scheduled tribe polyandrous male was 14.3 as compare to scheduled caste male polyandrous that as 12.85 the difference between

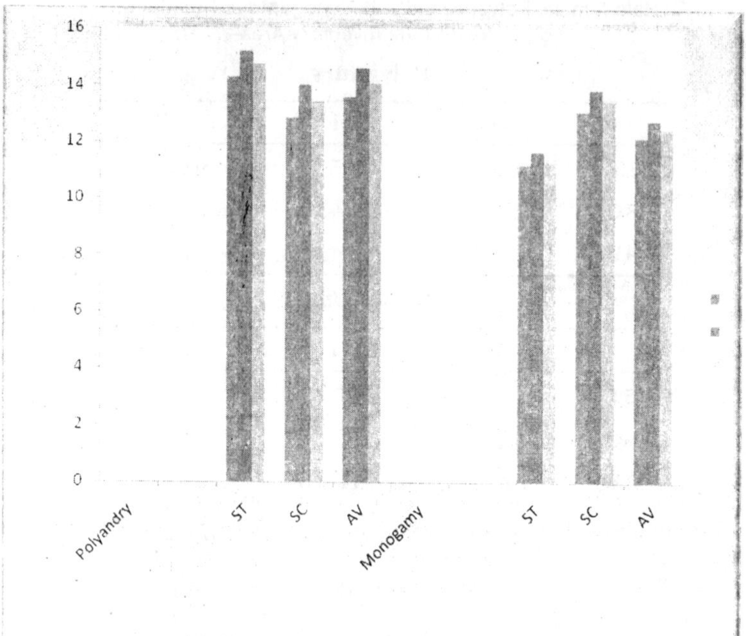

Figure 6.6.6.f: The Average scores of scheduled tribe and scheduled caste, male-female, polyandrous and monogamous population on polyandry attitude inventory-area-(A+B)

Notation: ST/SC = Scheduled Tribe/Caste; M/F = Male/Female; AV = Average scores two is (14.3-12.85)=1.45. So it quite clear that scheduled caste male has more causes and consequences than scheduled tribe male polyandrous. The average score of polyandrous female scheduled tribe was 15.2 as compare to scheduled caste female polyandrous that was 14.0. The difference between both is (15.2-14.0) = 1.2. So this difference shows major negligible difference between the two.

The average score of male polyandrous scheduled tribe as 14.3 and female scheduled tribe polyandrous as 15.2 so the difference between both is (15.2-14.3) = .9, so it quite clears that male polyandrous scheduled tribe more causes and consequences than female polyandrous scheduled tribe. The average score of scheduled caste male polyandrous as 12.85 and the average of female scheduled caste polyandrous as 14.0. The difference is (14.0-12.85) = 1.15 and it shows that male scheduled caste polyandrous has more causes and consequences than female scheduled caste polyandrous. The average score of scheduled tribe polyandrous male was 14.3 and average score of female scheduled caste polyandrous as 14.0. Now on comparison the difference came as (14.3-14.0) = .3. Thus there is slight difference between them and it shows that female scheduled caste polyandrous has more causes and consequences than male scheduled tribe polyandrous. Finally the average score of female scheduled tribe as 12.85 where as the average score of male polyandrous as 14.0. The difference between two is (15.2-12.85) = 2.35. Thus the same is shown that scheduled caste male polyandrous has more causes and consequences than scheduled tribe female.

The Average score of monogamous scheduled tribe on polyandry attitude Inventory was 11.425 where as the average score of scheduled caste monogamous people is 13.475. The difference between two is (14.425-13.475) = 2.05. Thus the picture is clear that the monogamous scheduled tribe has high causes and consequences than scheduled caste monogamous. The average score of monogamous male as 12.15 and of monogamous female was 12.75. Now on comparison the difference came as (12.75-12.15) = .6. Thus there is slight difference between them and monogamous male has high causes than female monogamous. The average score of scheduled tribe monogamous male was 11.2 as compare to scheduled caste male monogamous that as 13.1 and the difference came as (13.1-11.2) = 1.9. So it is very much clear that male scheduled caste has less causes and consequences than male scheduled tribe monogamous. The average score of female scheduled tribe as 11.65 and the average score of female monogamous scheduled caste as 13.85, the difference between both is (13.85-11.65) = 2.20. Showing scheduled caste monogamous female was less causes and consequences than female scheduled tribe.

The average score of male scheduled tribe monogamous as 11.2, the average score of female scheduled tribe monogamous as 11.65 and the difference between both is (11.65-11.2) = .4. So there is slight difference between them male scheduled tribe has more causes and consequences than female scheduled tribe. The average score of male monogamous scheduled caste was 13.1 and the average score of female monogamous scheduled caste as 13.85. The difference between the two is (13.85-13.1) = .75. Thus it is slowly that the monogamous female scheduled caste has less causes and consequences than male scheduled caste. The average score of male scheduled tribe monogamous as 11.2, the average score of scheduled caste female monogamous as 13.85 and the difference between both is (13.85-11.2) = 2.65. Thus the picture is so clear that female scheduled caste monogamous has less causes and consequences than male scheduled tribe monogamous. The average score of male scheduled caste monogamous as 13.1, the average score of female scheduled tribe as 11.65 and the difference between them is (13.1-11.65) = 1.45. Showing that male scheduled caste has less causes and consequences than female scheduled tribe.

The average score of polyandrous scheduled tribe as 14.75 where as average score of monogamous scheduled tribe as 11.425 and the difference is (14.75-11.425) = 3.325, it showing scheduled tribe monogamous has more causes and consequences than polyandrous scheduled tribe. The average score of polyandrous scheduled tribe as 14.75 and the average score of monogamous scheduled caste as 13.475 now difference came as (14.75-13.475) = 1.275. It shows monogamous scheduled caste has more causes and consequences. The average score of polyandrous scheduled caste was 13.425, the average score of monogamous scheduled tribe as 11.425 and the difference came as (13.425-11.425) = 2.0. Thus it show monogamous scheduled tribe has more causes and consequences. The average score of polyandrous scheduled caste as 13.425 and average score of monogamous scheduled caste as 13.475 and the difference between both is (13.475-13.425) = .05. Thus showing monogamous scheduled caste has less causes and consequences than polyandrous scheduled caste. The average score of polyandrous (male, female, scheduled caste, scheduled tribe) as 14.087, the average score of monogamous (male, female, scheduled caste, scheduled tribe) as 12.45 and the difference is (14.087-12.45) = 1.637. It shows that monogamous has more causes and consequences of polyandrous marriage.

6.7 STATE-ANXIETY AMONG TRIBAL POLYANDROUS AND MONOGAMOUS PEOPLE

In general, personality state is situational and traits have been described as relatively enduring individual differences among people in specifiable tendencies to perceive the world in a certain way and in disposition to react or behave in a specified manner with predictable regularity. Personality traits reflect individual differences in the frequency and intensity with which certain emotion states have been manifested and in the probability of occurrence of such states in frequency and intensity of an emotional state depends upon the strength of personality traits (A-trait) has been defined in terms of relatively stable individual differences in anxiety proneness, in contrast to trait, state anxiety is momentary or for short period of time. Personality state is situational.

Table 6.6.7: Average scores of Scheduled Tribe and Scheduled Caste, Male-female, polyandrous and monogamous People on states Anxiety-Inventory

Polyandry

	M	F	AV
ST	37.7	37.7	37.7
SC	45.4	48.2	46.8
AV	41.55	42.95	42.25

Monogamy

	M	F	AV
ST	33.0	36.1	34.55
SC	43.0	49.1	46.05
AV	38	42.6	40.30

Notation: ST/SC = Scheduled Tribal/Caste; M/F = Male/Female; AV = Average scores

From the table 6.6.7. It is quite clear that average score of polyandrous scheduled tribe people on state anxiety as 37.7 and average score of scheduled caste polyandrous as 46.8. The difference between the two, sets of score was (46.8-37.7) = 9.1. Thus it shows quite clear difference between them and polyandrous scheduled caste people has experienced more state anxiety problem as compared to polyandrous scheduled tribe counterpart. The average score of polyandrous male on state anxiety was 41.55 and whereas the polyandry female as 42.95. The difference between the two sets of score is (42.95-41.55) = 1.40. Thus it is clear that the female polyandrous has more anxiety than male scheduled tribe polyandrous. The average score of scheduled tribe polyandrous male was 37.7 as compared to scheduled caste male 45.4 and the difference between the two as (45.4-37.7) = 7.7. So it is very clear that scheduled caste male has experience more episodes of state anxiety than scheduled tribe male. The average score of polyandrous female scheduled tribe as 37.7 and scheduled caste female polyandrous as 48.2. The difference between both is (48.2-37.7) = 10.5. Thus the picture is so clear that the scheduled caste female has witnessed more state anxiety than scheduled tribe female polyandrous, counterpart.

Similarly, the average score of scheduled tribe polyandrous male as 37.7 and scheduled tribe polyandrous female as 37.7. As the difference came nil/zero so it shows equal anxiety between the two. The average score of scheduled caste male polyandrous was 45.4 and the average of female scheduled caste as 48.2. The difference is 48.2-45.4) = 2.8. So this difference shows that female scheduled caste polyandrous has high state anxiety than male scheduled caste polyandrous. The average score of scheduled tribe polyandrous male as 37.7 and the scheduled caste polyandrous female as 48.2. Now on comparison the difference came out as (48.2-37.7)=10.5. Thus there is clear difference between them and it shows that the male scheduled tribe has low state anxiety than their counterpart. Finally the average score of scheduled tribe Polyandrous female as 37.7 whereas the average score of scheduled caste polyandry male as 45.4. The difference between the two is (45.4-37.7)=7.7. Thus it shows that there is quite clear difference between them and the scheduled caste polyandrous male has high prevalence of anxiety than female scheduled tribe, counterpart

The average score of scheduled tribe monogamous population on state anxiety as 34.55 whereas the average score of scheduled caste monogamous as 46.05. The difference between two is (46.05-34.55)=11.40. Thus the picture is so clear that the monogamous scheduled tribe people have low state anxiety than monogamous scheduled caste people. In the same manner, the average score of monogamous male on state anxiety aspects as 3.8 and monogamy female as 42.6. Now on comparison the difference came as (42.6-38) = 4.6. Thus there is clear difference between them and the monogamous female has high anxiety than male monogamous. Average score of scheduled tribe monogamous was 33.0 as compared to scheduled caste male was 43.0. The difference is (43.0-33.0) = 10 showing that male scheduled tribe monogamous has less state anxiety than scheduled caste male. The average score of female scheduled tribe as 36.1 and average score of female scheduled caste as 49.1. The difference is (49.1-36.1) = 13.0. So picture is clear that female scheduled tribe monogamous has very less state anxiety than scheduled caste female.

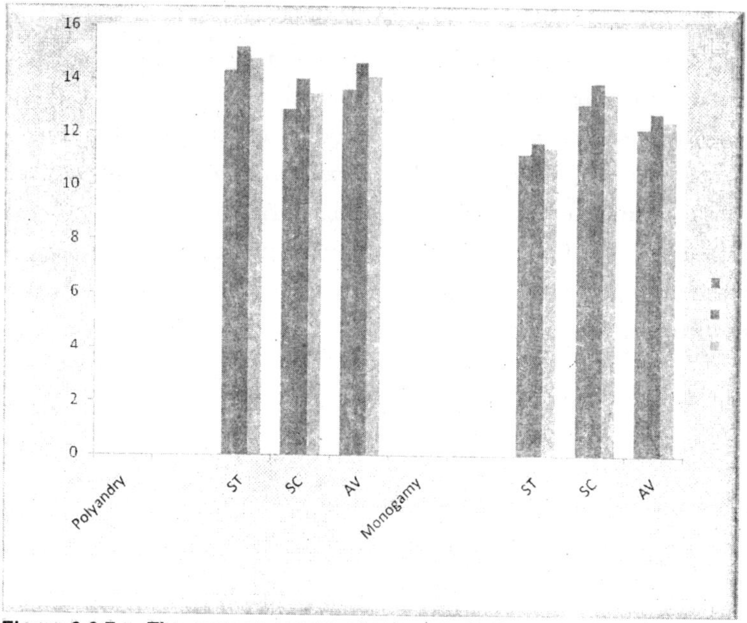

Figure 6.6.7.g: The average scores of scheduled tribe and scheduled caste, male-female, polyandrous and monogamous People on state anxiety-inventory

Notation: ST/SC = Scheduled Tribe/Caste; M/F = Male/Female; AV = Average scores

Finally the average score of scheduled tribe monogamous male as 33.0. The average score of scheduled tribe monogamous female as 36.1 and the difference is (36.1-33.0)= 3.1. Thus there is clear difference between them and thus scheduled tribe female has been suffering with more state anxiety than male scheduled tribe. The average score of scheduled caste monogamous male as 43.0 and average score of scheduled caste monogamous female as 49.1. The difference is (49.1-43.0) = 6.1. So it is quite clear that female scheduled caste has very high state anxiety than scheduled caste male. The average score of male monogamous scheduled tribe as 33.0 whereas the average score of the scheduled caste monogamous female as 49.1, the difference is (49.1-33.0) = 16.1. So it is very much clear that female scheduled caste monogamous has reported high state anxiety. Furthermore, the average score of scheduled caste monogamous male as 43.0 and average score of scheduled tribe monogamous female as 36.1, the difference is (43.0-36.1) = 6.9. So it is clear that the scheduled tribe female has reported low state anxiety than scheduled caste male monogamous counterpart.

CONCLUSION

The objective of the present study was to study the psychological wellbeing among polyandrous and monogamous socially disadvantaged tribal people of Kinnaur in Himachal Pradesh. In this study an attempt was made to explore the psychological wellbeing among the scheduled tribe and scheduled castes population including men and women from polyandrous and monogamous population of Kinnaur district. The study is pioneer in nature because the social institution and their racial system have not been studied so far in a more comprehensive way in the tribal areas those perhaps are directly associated with ones wellbeing. It is generally believed that there do exist only tribal population in tribal areas whereas the fact is that the scheduled caste has remained introvert and dormant by in the eyes of the researchers especially in the academician, policy makers and politician. There do excised complex socio-cultural heterogeneity and caste system and marital style among the scheduled tribes of District Kinnaur. Therefore the specific objectives in the study were to explore the physical and psychological well-being, polyandry attitude and anxiety among the Scheduled tribe and Scheduled caste polyandrous and monogamous men and of Kinnaur District in Himachal Pradesh. For accomplishing the objective the study was accomplished in different areas of district

Kinnaur which is situated on National Highway-22 in Himachal Pradesh (India). More specifically, the present study had been conducted in villages Sapni, Sangala, Monne, Aakpa and Kalpa. The villages fall in Nichar, Kalpa and Pooh Blocks of the district Kinnaur. These villages are approximately 260 kilometer from Shimla and 100 kilometer from Rampur Bushahar approachable through public and private buses. In the present study a sample of N=80 subjects was selected from aforesaid villages of district Kinnaur. The sample selected in the present study was above 25 year of age. The sample of i.e N = 80 subjects were divided into two sub–groups based on their marital status i.e. from Polyandry (N=40) and Monogamy (N=40) subjects were selected. These subjects were again divided into two comparable half based on their Gender that comprises of N=20 in male groups as well as another N=20 in females groups. Finally, the subjects were further sub-divided into two comparable halves based on their caste that comprises of N=10 in Scheduled Tribe groups as well as another N=10 in Scheduled Caste groups. In all, there were eight groups with N=10 subjects in each that comprise of complete sample of 80 subjects of aforesaid characteristics. A 2 × 2 × 2 factorial design was used. Both qualitative and quantitative analysis was used. In qualitative measure observation, Interview, Fuzzy Cognitive Mapping and photography were used. In quantitative measures PGI Health Questionnaire, Polyandry Attitude Inventory and State Anxiety Inventory were used. The quantitative measures are self-report in nature that will be administered by the researcher herself by asking open ended question. Since the researcher herself hails from Kinnaur district it helped her a lot in carrying out numerical for this manuscript. Beside this the researchers did her pilot study and asked various people about polyandry system in this district. Various

Gram Pradhan, educated youths as well as elderly people were asked to know deeply about polyandrous people in the concerned villages. The researcher entered in the particular village, established warmth atmospheres by addressing the objective of the study and collected appropriate information's. Analysis of variance was used and the results are as follows:-

Three way analysis of variance was performed on the PGI-Health questionnaire N-1. It specifically deals with the physical aspects of the neuroticism. The main effect of marriage in physical aspects of neuroticism was found $F(1, 72) = .18$, $p > .05$ statistically non- significant. But the main effect of sex was found $F(1, 72) = 3.08$, $p < .01$ as statistically significant. Further the main effect of caste was found $F(1, 72) = 7.06$, $p < .01$ as statistically significant. The two way interaction between marriage and sex (MA × SX) was found $F(1, 72) = .01$, $P > .05$ non significant. Similarly the two way interaction between marriage and caste (MA × CA) was found $F(1, 72) = 2.54$, $P > .05$ as statistically non-significant. Further the two way interaction between sex and caste was found $F(1, 72) = .48$, $p > .05$ non significant. Finally the three way interaction between marriage sex and caste (MA × SX × CA) was also found $F(1, 72) = 0.07$, $p > 0.5$ as statistically non-significant.

Three way analysis of variance was performed on the PGI-Health questionnaire N-2. It specifically deals with the psychological aspects of the neuroticism. The main effect of marriage was found $F(1, 72) = .12$, $p > .05$ as statistically non-significant. But the main effect of sex was found $F(1, 72) = 6.18$, $p < .01$ as statistically significant. Finally r the main effect of caste was found $F(1, 72) = 26.21$, $p < .001$ as highly statistically significant. The two way interaction between marriage and sex (MA × SX) was found $F(1, 72)$

=1.1, $p>.05$ non significant. Similarly the two way interaction between marriage and caste (MA × CA) was found $F(1, 72) = 1.7$, $p>.05$ as statistically non-significant. Further the two way interaction between sex and caste was found $F(1, 72) = 2.37$, $p>.05$ as statistically non significant. Finally the three way interaction between marriage sex and caste (MA × SX × CA) was also found $F(1, 72) = 1.20$, $p>0.5$ as statistically non-significant.

Three way analysis of variance was performed on the PGI-Health questioner total scores of both areas. It specifically deals with both psychological and physical aspects of the neuroticism. In both the aspects main effect of marriage was found $F(1, 72) = 1.7$, $p>.05$ non significant. But the main effect of sex was also found $F(1, 72) = 5.30$, $p<.01$ as statistically significant. Further the main effect of caste was found $F(1, 72) = 18.02$, $p<.001$ as highly statistically significant. The two way interaction between marriage and sex (MA × SX) was found $F(1, 72) = .31$, $P>.05$ non significant. Similarly two way interaction between marriage and caste (MA × CA) was found $F(1, 72) = 2.41$, $P>.05$ as statistically non-significant. The two way interaction between sex and caste was found $F(1, 72) = 1.49$, $p>.05$ non significant. Finally the three way interaction between marriage sex and caste (MA × SX × CA) was also found $F(1, 72) = 0.56$, $p>0.5$ as statistically non-significant.

Three way analysis of variance was performed on the Polyandry Attitude Inventory Area-1. It specifically deals with causes of Polyandry and polygamy people. The main effect of marriage was found $F(1, 72) = 3.433$, $p<.05$ as statistically significant. But the main effect of sex was found $F(1, 72) = 4.73$, $p<.05$ as statistically significant. Further the main effect of caste was found $F(1, 72) = 2.4$, $p>.05$ as statistically Non-significant. The two way interaction between marriage and sex (MA × SX) was found F (1, 72) =.649, p>.05 non significant. Similarly two way interaction between marriage and caste (MA × CA) was found F (1, 72) =6.65, p<.01 as statistically significant. Further the two way interaction between sex and caste was found F (1, 72) =.23, p>.05 non significant. Finally the three way interaction between marriage sex and caste (MA × SX × CA) was also found F (1, 72) =.006, p>0.5 as statistically non-significant.

Three way analysis of variance was performed on the Polyandry Attitude Inventory Area-2. It specifically deals with the consequences of polyandry on well-being of people. In both the aspects main effect of marriage was found F (1, 72) =9.2, p<.01 as statistically significant. But the main effect of sex was found F (1, 72) =.15, p>.05 as statistically non-significant. Further the main effect of caste was also found F (1, 72) =.10, p>.05 as statistically non-significant. The two way interaction between marriage and sex (MA × SX) was found F (1, 72) =.01, p>.05 non significant. Similarly two way interaction between marriage and caste (MA × CA) was found F (1, 72) =6.42, p<.051 as statistically significant. Further the two way interaction between sex and caste was found F (1, 72) =.00, p>.05 non significant. Finally the three way interaction between marriage sex and caste (MA × SX × CA) was also found F (1, 72) =.01, p>0.5 as statistically non-significant.

Three way analysis of variance was performed on the Polyandry Attitude Inventory both causes and consequences. It specifically deals with the cause and consciences of Polyandry and polygamy people. In both the aspects main effect of marriage was found F (1, 72) =8.78, p<.01 as statistically significant. But the main effect of sex was found

F (1, 72) =2.16, p<.01 as statistically non- significant. Further the main effect of caste was found F (1, 72) =.430 p>.05 as highly statistically non-significant. The two way interaction between marriage and sex (MA × SX) was found F (1, 72) =.15, p>.05 non significant. Similarly two way interaction between marriage and caste (MA × CA) was found F (1, 72) =9.32, p<.01 as statistically significant. Further the two way interaction between sex and caste was found F (1, 72) =.06, p>.05 non significant. Finally the three way interaction between marriage sex and caste (MA × SX × CA) was also found F (1, 72) =.01, p>0.5 as statistically non-significant.

Three way analysis of variance was performed on the state anxiety. It specifically deals with the anxiety of people. In both the aspects main effect of marriage was found F (1, 72) =.53, p>.05 non significant. But the main effect of sex was found F (1, 72) =1.27, p>.05 as statistically non-significant. But the main effect of caste was found F (1, 72) =15.27, p<.001 as highly statistically significant. The two way interaction between marriage and sex (MA × SX) was found F (1, 72) =.35, P>.05 non significant. Similarly two way interaction between marriage and caste (MA × CA) was found F (1, 72) =.21, P>.05 as statistically non-significant. Further the two way interaction between sex and caste was found F (1, 72) =.29, p>.05 non-significant. Finally the three way interaction between marriage sex and caste (MA × SX × CA) was also found F (1, 72) =0.00, p>0.5 as statistically non-significant. Three way analysis of variance was performed on PGI Health questionnaire- physical aspects and the result revealed that the main effect of sex F (1, 72) =3.08, p<.01 and caste was found F (1, 72) =7.06, p<.01 as statistically significant. Other interactions emerged as statically non-significant.

SUGGESTIONS

1. The jail inmates be dealt with reform and humanitarian approach instead of fire-fighting approach. An opportunity be provided for basking light, safe shelter, appropriate blanket, cots, followed with sanitation and drainage facilities. Hygienic environment be provided to them. The jail inmates be provided three time healthy and nutritious food.
2. Recreational facilities be increased that include games, television, video, radio and email etc. Appropriate material be provided to them. Picnic facilities be also provided once in a year to the jail inmates.
3. The jail inmates be encouraged to pursue education, to impart professional education in order to enhance their capabilities. Library facilities be enhanced and new news bulletin, news papers, magazine, journals, books and thesis may be provided to them. Beside this religious books, and books related to psychology especially to the moral and personality development be provided to them. Workshop, seminar and conference be started so as to modify their thought process.
4. The inmates be provided job in all the sectors after completing their term. Risk intervention team and

high risk alert team may launch treatment package that in turn may prove effective for the same. Further prisoner support schemes and case management team may help a lot. Further there is a need to increase role of fellow prisoners and extension peer support.

5. New system for monitoring punishment and adjudicating grievance, prison routine, safety issues ad support services be provided to them. Prisoner handbook may be provided to them. Constructive activity within the prison be started. Support from prison staffs and probation officer be provided. Hopes and plans for future be provided.
6. Family support and visit be provided to them. Separate guest room be built near the jail where the jail inmates can stay with their family members such as spouse.
7. Inter-departmental communication be started. The staffs that are professionally trained and valued be deputed.
8. Mental health services should be integrated with Health Service Executives. Yoga, meditation, art and music such as Bhajan Kirtan therapy be encouraged through instructor. Thus diversion into non-forensic (outside) mental health services is needed to them.
9. Proper maintenance of Medical Record be kept. Regular checkup, treatment to the civil hospital even in minor diseases
10. Exchange of information with community care providers and residential care provider.Development of non-pharmacological intervention. Self-harm/suicide support and assessment program.

11. There is a dire need of Psychologists especially a counselor in order to impart counselling as well as to see the impact of treatment in order to know the changes in behaviour and thought processes. The counselor should appraise their goal, aim and objectives of the life by appropriate behavioural modification techniques.
12. The prisoner be rewarded and reinforced for their good behaviour. The term of jail should be short not lengthy because the criminal behaviour is situation specific and momentary. Once commiting the crime the criminal repent throughout the life.
13. Although some short term prisonment is necessary yet it should be converted into finance after its commitment. If the prisoner again and again (2-3 time) commits the same crime then there should be a provision of capital punishment.
14. The prison duration be short in which the inmates be dealt with humanitarian approach by forgetting fire fighting approach.
15. There should be a complaint boxes that should be tackled only by the supreme authority not by the jail staffs. In this respects action be also taken to the jail staffs.

REFERENCES

Aiyar, S. A. A. (2010). Nutrition is a bigger problem than hunger. *The Times of India, September,* 2010. Chandigarh: HT Press.

Joyce, A. K. (1997). Ethnicity, gender and identity among second-generation Arab Americans: Growing up Arabic in America. *Ph.D. Thesis,* Wayne State University, Detroit.

Al-krenawi, A., Graham, J. R., & Slonim-nevo, V. (2002).). Mental health aspects of Arab-Israeli Adolescents from polygamous versus monogamous families. *The Journal of Social Psychology, 142,* 4, 446 - 460

Andrews, F., & Withey, S.(1974). Developing measures of perceived life quality. *Social Indicators Research,* 1, 1–26.

Antognoli-Toland, Paula Lucia. (2000). Parent-child relationship, family structure, and loneliness among adolescents. *Ph.D. Thesis,* Texas Woman's university.

Ancell, S. (2009). *Marriage good news for mental health, but separation and divorce have negative impacts.* Newzealand: Otagio University.

August-Prudhomme, N. R. (1997). The relationship between family functioning and female adolescent sexual behavior. *Ph.D. Thesis,* Louisiana State University Medical Ctr. In New Orleans, S. Of Nursing.

Bandura, A. (1977). Analysis of self-efficacy theory of behavioral change. *Psychological Review, 84,* 191-215.

Behrman, J.R., & Wolfe, B.L. (1984). More evidence on nutrition demand: income seems overrated and women's schooling underemphasized. *Journal of Development Economics 14,* 105 128.

Benzeval, M., Judge, K. & Whitehead, M. (1995). *Tackling inequalities in health.* London: Kings Fund.

Bollen, K. (1989). *Structural equations with latent variables.* New York: Wiley.

Bradley, R. (1989). Home measurement of maternal responsiveness. In: M.H. Bornstein (Ed.), *Maternal responsiveness: Characteristics and consequences (pp. 63-74).* New York: Jossey-Bass, 63-74.

Brody, G. H., Stoneman, Z., Flor, D. & McCrary, C. (1994). Religion's role in organizing family relationships - family process in rural, 2-parent African-American families. *Journal of Marriage & the Family, 56,* 878-888.

Bowling, A. (1995a).What things are important in people's lives? A survey of the public's judgments to inform scales of health related quality of life. *Social Science & Medicine, 41, 1447-* 1462.

Bowling, A, C. (1995). *Measuring disease: A review of disease specific quality of life measurement scales.* Buckingham U K: Open University Press.

Bowling, A. (1995b). The most important things in life. Comparisons between older and younger population age groups by gender. Results from a national survey of the public's judgments. *International Journal of Health Sciences, 6,* 169-175.

Bowling, A. (1995b). What things are important in people's lives? A survey of the public's judgments to inform scales of health related quality of life. *Social Science & Medicine, 41, 1447-* 1462.

Brody, G. H., Zolinda, S. & Douglas, F. (1996). Parental religiosity, family processes, and youth competence in rural, two-parent African-American. *Journal of Marriage & the Family, 89,* 1878-1988.

Bybee, Linda, Millar, Jocelyn , Paine, Timothy , Campbell, Kathleen et al, (2005).Effects of single versus multiple mates: monogamy result in increased fecundity for the beetle phoracantha semipunctata. Department of Entomology, University of California, Riverside, CA. *Journal of Insect Behavior,18*(4), 513-527.

Calvert, W. J. (1997). Protective factors within the family, and their role in fostering resiliency in African American adolescents. *Journal of Cultural Diversity, 4,* 110-117.

Campbell, A., Phil, C. & Willard, R. (1976). *The quality of American life.* New York: Russell Sage Foundation.

Campbell, A., Converse, P. L., & Roger, W. L. (1976). *The quality of American life: perception, evolution and satisfaction.* New York: Sage.

Campbell, A., Converse, P. & Rogers, W.(1976). *The quality of American life.* New York: Russell Sage Foundation,

Crook, J., & Osmaston, H. (Eds.) (1994). *Himalayan Buddhist villages.* Delhi: Moti Lal Banarsidass.

Cummins, R. A. (1997). Assessing quality of life. In R. I. Brown (Ed.), *Quality of life for people with disabilities-second edition: Model research and practice (pp. 116-150).* Cheltenham: Stanley Thomas.

Dhawan, H. & Ghosh, D. (2010). Centre issues alert on sex tourism. *The Times of India, September 10,* 2010.

Diener, E. (1984). Subjective wellbeing. *Psychological Bulletin, 95,* 542-575.

Diener, Ed. (1995). A value-based index for measuring National quality of life'. *Social Indicators Research,* 36, 107-127.

Dodd, David S. (1997). Missionary family life: Resources and strategies for dealing with stress. *Ph.D. Thesis,* Rosemead School of Psychology Biola University.

Eagly, A., & Chaiken, S. (1995). Attitude strength, attitude structure and resistance to change. In R. Petty and J. Kosnik (Eds.), *Attitude Strength* (pp. 413-432). Mahwah, NJ: Erlbaum.

Elbedour S., Bart, W., & Hektner, J. (2007). The relationship between monogamous/polygamous family structure and the mental health of Bedouin Arab adolescents. *Journal of Adolescence, 30,* 213-230

Elder, Glen H. Jr. (1962). Structural variations in the child rearing relationship. *Sociometry, 25,* 241-262.

Elder, G. H., V. King, & R. D. Conger. (1996). Attachment to place and migration prospects: A developmental perspective. *Journal of Research on Adolescence* , 6, 397-425.

Elwin,V. (1958). *Tribal women, in women of India. Also in Devaki Jain (Ed.) Indian Women.* New Delhi: Publication Division, Ministry of Information and Broadcasting, 1975.

Embry, Jessie L. & Martha S. Bradley. (1985). Mothers and daughters in polygamy. *Dialogue, 18,* 98-107.

Engle, P.L. (1992). Care and nutrition. *ICN Theme Paper.* Report prepared for UNICEF.

Erich, Stephen & Patrick Leung. (1998). Factors contributing to family functioning of adoptive children with special needs: A long term outcome analysis. *Children and Youth Services Review, 20,* 135-150.

Erci, Behice & Ergin, Raziye. (2005). Women's satisfaction with their marriage in Turkey. public health nursing department school of nursing, Ataturk University,Eezurum,Turky. *Marriage & Family Review, 37 (3), 117-133.*

Evans-Pritchard, E.E. (1951). *Kinship and marriage among the Nuer.* Oxford: clarendon Press.

Farber, B. (1964). *Family: Organization and interaction.* San Francisco: Chandler Publishing Co.

Goode, W. J. (1967). *Family disorganization.* In R. K. Merton & R Nisbet (Eds.), Contemporary social problems. New York: Harcourt, Brace, & Johanvich Publication.

Goldscheider, Calvin and Frances Kobrin Goldscheider. (1987). Moving out and marriage: What do young adults expect? *American Sociological Review, 52,* 278-285.

Grant, P.J. (1984). In: *The state of the world children.* New York: Oxford University Press for UNICEF, 3.

Greeff, A. P. & le Roux, M. C. (1999). Parents and adolescents perceptions of a strong family." *Psychological Reports, 84,* 1219-1224.

Hammond, J. A., Cole B. S. & Beck, S. H. (1993). Religious heritage and teenage marriage. *Review of Religious Research, 35,* 117-133.

Handerson, H.(2002). A system approach: Calvert-Henderson quality of life indicators. *www.hastehanderson.com.*

Hanson, Shirley M. (1986). Healthy single parent families. *Journal of Applied Family &Child Studies, 35,* 125-132.

Hartta, Y., Zinta, R. L., & Tiwari, A. K. (2009). Appraisal of HIV-AIDS consciousness among college youths in Himachal Pradesh. *Social Change, 39 (3),* 406-428.

Heller, P., & Drake, W. (1979). Malnutrition, child morbidity, and the family decision process. *Journal of Development Economics 6, 21, 3-235.*

Helsing, K. J., & Szklo, M. (1981). Mortality after bereavement. *American Journal of Epidemiology, 114,* 41-52.

Herrick, Susan Carol. (1981). *Sibling violence: does piety make a difference?* Paper presented at Association for the Sociology of Religion (ASR).

Hills, J. (1995*). Inquiry into income and wealth, 2.* Joseph. Rowntree Foundation, York.

Hilliard, Donnie Ray. (1996). Qualities of successful father-child relationships. *Ph.D. Thesis,* The University of Alabama.

Hodson, T.C. (1975). *The meithesis.* Delhi: B.R. Publishing Corporation.

Holmes, L. (2004). Partnership and mental health over time. *Journal of Epidemiology &Community Health 2004:58*; 53-8

Hornback, K., & Shaw, R. (1973). Towards a quantitative measure of the quality of life. In US Environmental Protection Agency (Ed.), *The Quality of Life Concept.* Government Printing Office, Washington, DC (pp. 13–24).

House, J. S. & Kahn, L. R. (1985). Measures and concepts of social support. In Cohen, S. & Syme S. L. (Eds.), *Social support and health.* London: Academic Press.

Hughes, R. S. (1999). An investigation of coping skills of parents of children with disabilities: implications for service providers.

Education and Training in Mental Retardation and Developmental Disabilities, 34, 271-280.

Hurst, Nancy. (2005). Marriage that promote growth. cornerstone counseling center, Edmonton, AB, Canada. *Marriage & Family Review, 37*(3), 47-71.

Ingoldsby, B. B. (2001). The hutterite family in transition. *Journal of Comparative Family Studies,* 32, 377-381.

Jankowiak, W., & M. Diderich, M. (2000). "Sibling solidarity in a polygamous community in the USA: unpacking inclusive fitness. *Evolution and Human Behavior, 21,* 125-139.

Johnson, K., Bryant. D. D., Collins, D. A., Noe, T. D., Strader, T. N., & Berbaum, M. (1998). Preventing and reducing alcohol and other drug use among high-risk youths by increasing family resilience. *Social Work, 43,* 297-308.

Josephson, Allan M. 1993. "The Interactional Problems of Christian Families and Their Relationship to Developmental Psychopathology: Implications for Treatment." *Journal of Psychology and Christianity, 12,* 312-328.

Kapadia, K.M (1966). *Marriage and family in India, (Fourth Edition).* New York: Oxford University Press.

Kour, R. (2010). The marriage roller coaster. *The Times of India 25 Augus* . Chandigarh: HT Press.

Keysar, A. & Kosmin, B. A. (1995). The impact of religious identification on differences in educational-attainment among American women in 1990. *Journal For the Scientific Study of Religion, 34,* 49-62.

Kishwar, M. (2010). Women can end this war. *Hindustan Times, 13 June, 2010.*

LeVine, P.M. Miller, & West, M.M. eds. (1987). *Parental behaviors in diverse* societies. *New Directions for Child Development, No. 40.* San Francisco: Jossey/Bass, 99-122.

Levine, Nancy E. 1988. *The dynamics of polyandry: kinship, domesticity and population on the Tibetan border. Chicago: University of Chicago Press. 309 . ISBN 0-226475697.*

Raha, M.K. (1991). Polyandry in India: Retrospect and prospect. *Man in India, 7* (1), (Special), 163-181.

Raha, M.K. & Coomar, P.C. (Eds.) (1987). *Polyandry in India.* Delhi: Gian Publishing House.

Raha, M.K. & Coomar, P.C. (Eds) (1988). The Kinnaur polyandry and the allocation of paternity. *Man and Life, 14* (2&4).

Ramey, Timothy R. (1996). The development of a mentor ministry to train the brotherhood of the barraque street missionary Baptist church. *Thesis, Midwestern Baptist Theological Seminary,* Kansas City.

Ramu, G. N. (1977). *Family and caste in urban India*: A case study. New Delhi: Vikas Publishing House Pvt. Ltd.

Rapley, M. (2003). *Quality of life research.* London: Sage.

Richter, J., Schwarz, M., & Bauer, B. (2008). Personality characteristics determine health-related quality of life as an outcome indicator of geriatric inpatient ehabilitation. current gerontology & geriatric research. *Published online doi: 10.1155/ 2008/474618.*

Ricks, S. S. (1987). Antecedents of Adolescent Satisfaction with Family Life. *Ph.D. Thesis,* Brigham Young University, Provo.

Ross, A. D. (1960). *The Hindu family in its urban setting.* Toranto: Toronto University Press.

Ross, A. D. (1960). *The Hindu family in its urban setting,* Toronto: Toronto University Press.

Royal Anthropological Institute of Great Britain and Ireland.

Ryan, D., Barbara Dooley, B & Benson, C. (2008). Theoretical Perspectives on Post Migration Adaptation and Psychological Well-Being among Refugees: Towards a Resource-Based Model. *Journal of Refugee Studies, 10*

Saito, Yasuhiko. (1993). Trends and differentials in marital and family formation preferences of American youth, 1967-1989. *Ph.D. Thesis,* University of Southern California,· Los Angeles.

Samadarshini (2010). Chasing mental ghosts. *Times of India, September 29,* 2010. Chandigarh: HT Press.

Schoellmann, Edward R. (1990). A comparative study of paternal nurturance as experienced by select categories of early adolescent children. *Ph.D. Thesis*, Texas Southern University.

Segraves R. T (1980). Marriage and mental health. *J Sex Marital Ther. Fall, 6(3)*:187-98.

Semans, M. P. & Fish, L. S. (2000). Dissecting life with a jewish scalpel: a qualitative analysis of Jewish-centered family life. *Family Process*, 39,121-139.

Sen, A. (1993). *Capability and well-being.* In M. C. Nussbaum & A. Sen (Eds.), *The quality of life.* Britain: Oxford.

Sharma, A., & Zinta, R. L. (2008). Alcoholism: A case study. *Indian Journal of Psychology & Mental Health, 1(2),* 195-214.

Singh, S. H. & Zinta, R. L. (2008). DHAT Syndrome: A case study. *Journal of Community Guidance & Research, 25* (2), 173-184.

Sonpar, shobna. (2005). Marriage in India: clinical issues. contemporary family therapy: *An International Journal, 27(3),* 301-313

Smetana, J. G. (2000). Middle-class African American adolescents and parents conceptions of parental authority and parenting practices: A Longitudinal Investigation." *Child Development, 71,* 1672-1686.

Spielberger, C. D., Sharma, S., & Singh, M. (1973). Development of the Hindi edition of the state-trait anxiety inventory. *Indian Journal of Psychology, 18,* 11-20.

Steinberg, L. (2001). We know some things: parent-adolescent relationships in retrospect and prospect. *Journal of Research on Adolescence, 11,* 1-19.

Steven, S. & Ross, E. J. (1998). Marital status and happiness: a 17-nation study. *Journal of Marriage and the Family Subject: Family and marriage*

Stillwell, Peggy Taylor. (1988). Family stress, coping, and resources as perceived by adolescents in nuclear, single parent, and remarried families. *Ph.D. Thesis*, The Florida State University, Tallahassee.

Super, C, Beiser, A., Guldan, G., Ahmed, N., Zeitlin, J., Ahmed, M., & Sockalingam, S. (1988). *A behavioral study of positive deviance in young child nutrition and health in rurjal Bangladesh,* Report to the Officer of Health and the Asia and Near East Bureau of the Agency for International Development, Baltimore, Md.

Tamini & Kahrazei, (2010). Gernal health and life satisfaction of student in polygamy and monogamy families. university of Sistan and Baluchestan, Zahedan, Iran. *Journal of the Indian Academy of Applied Psychology, 36, 2,* 307-310.

Tiwari, A.K. & Zinta, R. L. (2007). Hydropower generation and economic development: A study of NJPC project in Himachal Pradesh. *Research Journal Social Sciences, 15,* 2, 136-148.

Vaillant, G. E. (2003). Mental health. *American Journal of Psychiatry, 160,* 1373-1384.

Vakalahi, H. F. (2001). Adolescent substance use and family-based risk and protective factors: a literature review." *Journal of Drug Education ,31,* 29-46.

Verma, S. K., Wig, N.M., & Pershad, D. (1985). *Manual for PGI health questionnaire N-1.* Agra: National Psychological Corporation.

Westmoreland, Cheri Lynn. (1996). Faith in action: A descriptive case study of project impact, a comprehensive juvenile diversion program sponsored by an African-American church. *Ph.d. Thesis,* University of Cincinnati.

Wilcock, R. O. (1992). Adolescent influences on young adult religious family values. *Ph.D. Thesis,* Brigham Young University.

Wilcox, W. Bradford. (1999). *Religion and paternal involvement: product of religious commitment or American convention?* Paper presented at American Sociological Association (ASA).

Wingo, L.(1973). The quality of life: towards a macroeconomic definition. *Urban Studies, 10,* 3–18.

Wyse, L. (2010). Make friends, live longer. *Hindustan Times, September 23,* 2010. Chandigarh: HT Press.

Zeitlin, M., Ghassemi, H., & Mansour, M. (1990). *Positive deviance in child nutrition: with emphasis on psychosocial and behavioural aspects and implications for development*. Tokyo: United Nations University Press.

Zinta, R. L. (2006). Is guided mastery helpful in problem solving. *Psychological Studies, 51(4),* 323-325.

Zinta, R. L. (2006). Effect of self-efficacy, test-anxiety and short-term intervention on problem solving: A preliminary analysis. *Social Science International, 22,* 1, 3-8.

Zinta, R. L. & Tiwari, A. K. (2006). Impact of socio-economic status on performance among male and female secondary students of Himachal Hill State. *Nava Arthiki, 14,* 2, 1-7.

Zinta, R. L. & Tiwari, A. K. (2006). Relationship between self-efficacy, socio-economic status and performance in secondary students. *Research Journal Social Sciences, 14,* (1), 121-137.

Zinta, R. L., & Tiwari, A. K. (2007). Impact of guided mastery treatment on self-efficacy among test-anxious students. *Journal of Psychological Researches, 51(2*), 91-101.

Zinta, R. L. (2008). Socio-economic status as a panacea for mental health: A note on displaced families of Bhakra Dam. *Nava Arthiki, 16, 2A, 49-74.*

Zinta, R. L., & Tiwari, A. K. (2008). Psychological analysis on project affected and likely to be affected families of SJVN Ltd. of Himachal Pradesh. *A project Reported submitted to the Integrated Institute of Himalayan Studies, HPU Shimla*-171005. (Recommended for publication).

Zinta, R. L., & Tiwari, A. K. (2008). Socio-economic and infrastructure facilities for sustainable rural development: A factorial analysis on Himachal Pradesh. A project Reported submitted to the *Integrated Institute of Himalayan Studies,* HPU Shimla-171005. (Recommended for publication).

Zinta, R. L. (2009). Bharat mein uch shiksha ki dasha aur disha. *Sanchar Bulletin, 1(1),* 104-106.

Zinta, R. L. (2009). Different facets of quality of life among the populace of mounting India in 21st century. *Akhil Bhartiya Shodh Seminar on Iksavin sadi mein Bharat: Dasha avam disha,* organized by University Grant Commission, New Delhi. Dayanand Vedic Snatkotar Mahavidyalaya, Urai, Jalon, U. P.

Zinta, R. L. (2010). Dimensions of quality of life in the tribal areas of Kinnaur: A caste-wise analysis. *A Research Project proposal approved by Institute of Tribal Research and Studies, H. P. University, Summer Hill,* Shimla-171005.

Zinta, R. L., Manjul, M., & Kumar, D. (2010). Social status and emotional wellbeing among employed and unemployed tribal single women. *Nava Arthiki, 18(1A),* 1-27.

Zinta, R. L., Azad, Y., & Tiwari, A. K. (2010). Appraisal of cognitive functioning among rural and urban deprived and non-deprived school students. *Nava Arthiki, 18(1A),* 35-52.

Zinta, R. L., Verma, L. K., Thakur, R. (2010). Inquesting psychological wellbeing among the squatters of Shimla town. *Nava Arthiki, 18(1A),* 61-91.

Zinta, R. L. (2010). *Psychology manual.* New Delhi: H. G. Publications.

Zinta, R. L. (2010). *Performance among high and low self-efficacious students.* In P, Ramalingam (Ed.), Recent studies in school psychology. Delhi: Authorspress.

Zinta, R. L. & Kour, G. J. (2013). Military personnel psychology. New Delhi: Neha Publication.